The Geology of Snowdonia and Llŷn:

An Outline and Field Guide

The Geology of Snowdonia and Llŷn:

An Outline and Field Guide

Brinley Roberts

Adam Hilger Ltd, Bristol

Consultant Editor
Dr. Brian Daley
Department of Geology
Portsmouth Polytechnic

British Library Cataloguing in Publication Data

Roberts, Brinley
The Geology of Snowdonia and Llŷn.
1. Geology—Wales—Snowdonia 2. Geology—
Wales—Lleyn Peninsula
I. Title
554.29′25 QE262.S/
ISBN 0-85274-290-8

Cover photograph of Clogwyn du'r Arddu by K J Wilson.

Published by Adam Hilger Ltd, Techno House, Redcliffe Way, Bristol BS1 6NX.
The Adam Hilger book-publishing imprint is owned by The Institute of Physics.

Printed in Great Britain by J W Arrowsmith Ltd, Bristol BS3 2NT.

Preface

This field guide has been written mainly with the undergraduate in mind, but it is hoped that naturalists, amateur geologists and sixth-formers might also find it useful. In the case of the sixth-former, it may well be that, in order to get much out of it, explanation and amplification will need to be provided by the teacher; while the amateur and general naturalist may need to consult standard texts. Some indication of the level of the guide can be gained from the fact that it is assumed in the case of igneous petrology that the user knows what is meant by, for example, a eutaxitic texture or an igneous cumulate; that in the case of structural geology he is able to distinguish between slaty, crenulation and fracture cleavages and understands the significance of minor structures; that in the case of stratigraphy he is familiar with the series, stages and zones of the Cambrian and Ordovician and knows something of the fauna; that in the case of sedimentology he can recognise the common types of sole-markings, ripples and cross beds.

The guide is divided into two sections. The first section briefly describes the succession and structure of the area, whereas the second section consists of about 30 itineraries. The intineraries have been compiled for use by the individual or very small group, rather than for larger parties. This is particularly true of some of the Llŷn itineraries where parking space is often restricted, exposures are limited, and permission may be needed to gain access to enclosed ground. These restrictions apply less to Snowdonia, and here an effort has been made to direct attention away from the more popular and better known areas to areas which are equally rewarding geologically, but are perhaps less dramatic scenically. Indeed, some of the popular and more readily accessible exposures are beginning to show the scars inflicted by a succession of parties of ravaging geologists. The user of this guide is therefore strongly urged to view the hammer as the least important of his geological impedimenta, and to use it sparingly and thoughtfully—if at all. In many cases useful specimens can be picked up from screes; but why collect at all if the material eventually finds its way into the dustbin? The important tools are maps, compass, clinometer, notebook and pencil.

A book of this kind, although reflecting the author's interests and prejudices, is usually largely a compilation of the work of others, and this is no exception. Where authors have been heavily drawn upon, acknowledgment is made in the text and, in the case of excursion maps, in the captions. Excursions in Central Snowdonia make use of the IGS 1:25 000 Geological Special Sheet (Central Snowdonia), and it is also useful to refer to this sheet when reading the first section of this book.

Wherever possible, Welsh place names have been spelt as recommended in the *Gazetteer* of Davies (1958). However, the earlier spellings of names which entered the geological literature have not been changed. Thus the reader will see reference to the village of Tremadog but to the Tremadoc Slates; to the town of Conwy but the Conway Volcanic Group.

The itineraries of the second section of this guide cover the ground from west to east. Thus excursions in Llŷn are followed by those in Eifionydd, then Arfon into northern Ardudwy, and finally in Arllechwedd. Many of the excursions are on high and rugged ground, and some are in rather remote areas. Bear in mind that weather conditions can change rapidly, so the appropriate equipment is necessary. Most importantly, the user of the guide must be able to use map and compass to find his way safely about the hills even in thick mist and driving rain. Remember that Easter can be a particularly dangerous time for the inexperienced; snow and ice can turn what is in summer a pleasant, easy scramble, into an ice pitch of some severity. References are collected together in a final section.

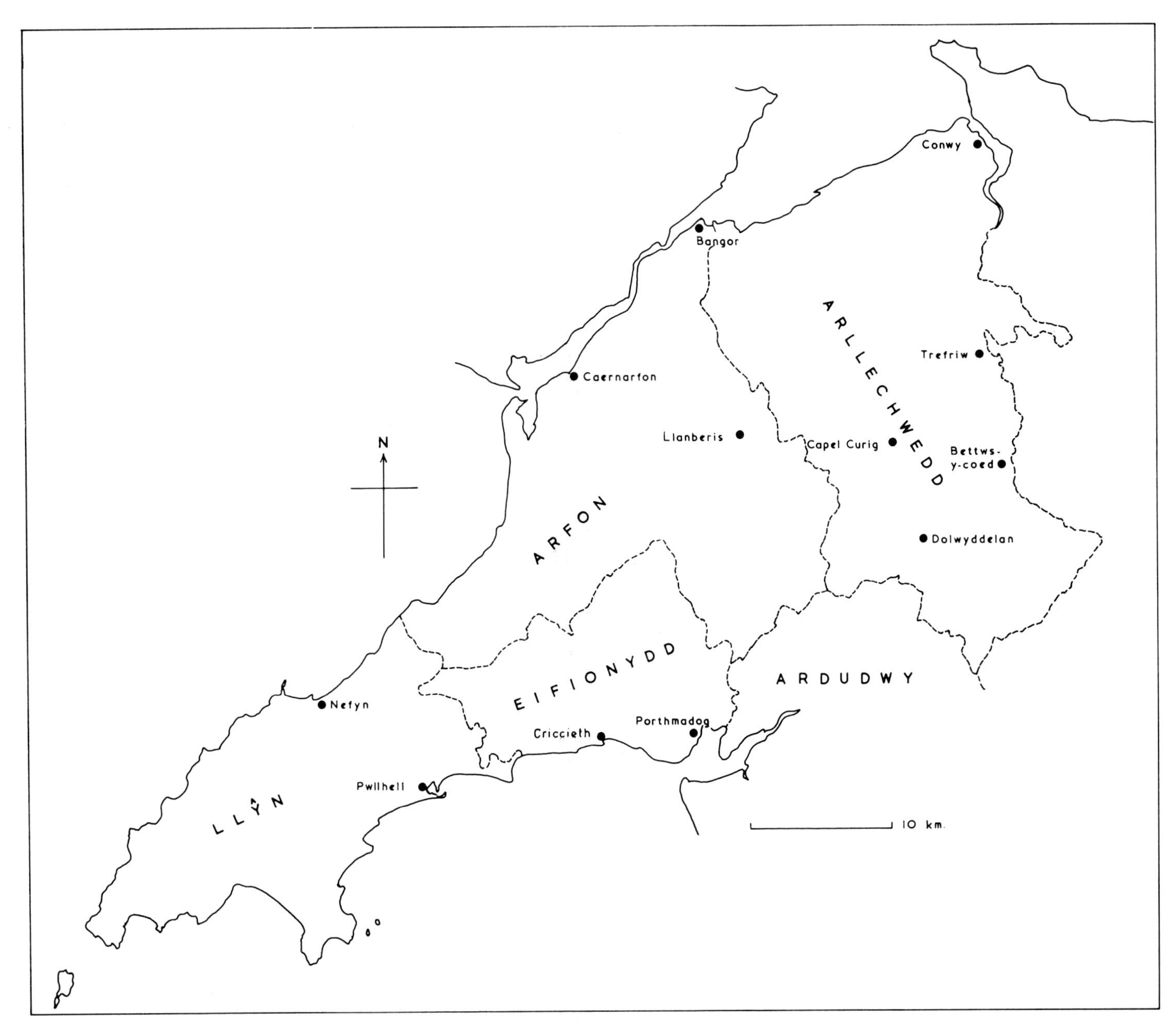

Figure 1. The districts of Snowdonia and Llŷn.

The unusual size and format of the guide is largely the result of the author's dissatisfaction with the format of many published guides. The large page size has been selected primarily in order to keep the excursion maps as large as possible. The author's ideal guide would not be bound in a conventional way, but would be of loose-leaf construction so that the pages and maps needed for a particular excursion could be temporarily removed from the guide and carried in the field in a map case. The flexible binding of this book is therefore a compromise, so that the guide can, if necessary, be rolled up and carried in the rucksack.

Finally, it is a pleasure to acknowledge the generous help of those authors who have allowed me to use their material, and to thank Janet Nelder for expertly typing and re-typing the manuscript. Dr D G Helm, Dr J Williams and Dr D Jenkins have each read and made comments on parts of the manuscript; but the guide would never have been completed had it not been for the kindness, hospitality and encouragement of John and Irene Williams.

Brinley Roberts

Contents

Part II Field Guide

Part I

An Outline of the Geology of Snowdonia and Llŷn

1. Geological History

The area to be described lies to the west of the Afon Conwy and north of the Vale of Ffestiniog. It includes, therefore, the northern portion of the Snowdonia National Park, together with the country lying to the north in Arfon and to the southwest in the peninsula of Llŷn.

The solid rocks range in age from Precambrian to Tertiary. The discontinuous but, in some places, thick mantle of Pleistocene deposits will not be described, despite the fact that evidence of glacial erosion and deposition is ubiquitous. This is because it was thought that an account of the Pleistocene geology of such an important area would be better left to a specialist in that field.

A very wide range of sedimentary, igneous and metamorphic rocks is present and, even for Great Britain, the variety in the geology is out of all proportion to the relatively small size of the area.

The rock series present may be summarised as shown in table 1.

This summary account is best read in conjunction with the geological map (end paper), and the IGS 1:25 000 Geological Special Sheet (Central Snowdonia) will be helpful. The area is covered by OS 1 : 50 000 sheets 115, 123 and 124.

Table 1. Summary of the rock series present. Note that (i) major unconformities only are indicated, and that (ii) formal stratigraphical divisions do not apply to the Precambrian.

Era	Period	Informal divisions	Series (where recognised)
Kainozoic	Tertiary	(Basic dykes only)	
Palaeozoic	Carboniferous	Upper Lower	
	Silurian	Lower	Llandovery
	Ordovician	Upper	Ashgill Caradoc Llandeilo (?)
		Lower	Llanvirn Arenig
	Cambrian	Upper	Tremadoc Merioneth
		Middle	St David's
		Lower	Comely
Precambrian		Arvonian Volcanic Group	
	Mona Complex	Gwyddel Felsitic Group Gwna Group	

1. The Mona Complex of Llŷn

The oldest rocks in the area are of Precambrian age and crop out in western Llŷn from Nefyn to Bardsey Island. Inland exposure is often poor but there are excellent, readily accessible coastal exposures. The rocks belong to the Mona Complex and have been described by Matley (1928), with contributions by Greenly, and by Shackleton (1954, 1956 and 1969). Matley recognised three main groups of rocks: a group of basic tuffs, pillow lavas and various sedimentary rocks, all in a very low grade of metamorphism; a group of medium-grade mica-schists; and finally a group of high-grade acid and basic gneisses. He regarded the schists as having been derived by medium-grade regional metamorphism from the group of pillow lavas and sedimentary rocks, but the gneisses were thought to be older and to represent the basement upon which the pillow lavas and sediments had accumulated. Shackleton (1954, 1956) supported Matley's view that there was a metamorphic transition from the pillow lavas and sediments into the schists, but he went on further to suggest a transition from the schists into the gneisses. The implication of this claim is that there is then no known basement upon which the rocks of the Mona Complex accumulated.

The type locality of the Mona Complex is Anglesey, where Shackleton separated the succession into seven members, but in Llŷn only the two youngest are exposed. The succession is shown in the table below.

Anglesey	Llŷn
Fydlyn Felsitic Beds Gwna Group	Gwyddel Felsitic Beds Gwna Group
Skerries Group New Harbour Group Rhoscolyn Beds Holyhead Quartzite South Stack Beds	*Not exposed*

The Gwna Group consists of a varied series of at least 1000 m of spilitic pillow lavas, basic tuffs, cherts, jaspers, limestones (some of which are oolitic), dolomites, quartzites, greywackes and pelites, whereas the Gwyddel Beds are about 100 m thick and are mainly fine-grained acid tuffs. The Gwna Group on Anglesey, and both the Gwna Group and the Gwyddel Group in Llŷn, are in the form of a mélange.

The formation now known as the mélange was first described by Matley (1913) on Bardsey Island, where it was described as a 'crush breccia' and 'crush conglomerate', and was regarded as a tectonic breccia. Greenly (1919) also described such a group found in Anglesey, where it was again interpreted as a tectonic breccia, but he referred to it as an 'autoclastic mélange'. The group, which is of regional extent, is the type example of a mélange.

Shackleton (1969), in describing the mélange, emphasised that the clasts range in size from the microscopic up to several kilometres across, and concluded that it is a subaqueous slide breccia or flow breccia. He compared the mélange with the olistostromes of the Appennines, and drew attention to the fact that slide breccias comparable with the Monian mélange occur characteristically in eugeosynclinal successions and must indicate tectonically induced instability.

Wood (1974) drew attention to the fact that in many places the mélange shows a convincing ghost stratigraphy, and that individual blocks of a particular lithology are separated by distances up to about 20 times the diameter of the blocks. Wood also pointed out that much of the mélange is still relatively flat-lying, and that to produce the disruption of the mélange by boudinage would necessitate a vertical principal compression acting normal to the bedding. Throughout the region, however, the tectonic structures above and below the mélange are steeply inclined. Wood suggested that much of the complication is the result of the fact that various member rock types were in different stages of lithification at the time of their initial movements: carbonate members were more lithified than quartzites which, in turn, were more lithified than pelites. He emphasised that in many places the ooliths of the limestone lack penetrative deformation and the matrix lacks evidence of ductile flow.

The pillow lavas of the Gwna Group are often beautifully preserved and form magnificent exposures at Porth Dinllaen and Dinas Bach. Shackleton (1956) has used the shape of the individual pillows to determine the direction of facing: pillows have a tendency to develop bun-shaped tops and indented bottoms. The lavas are spilitic, in that the plagioclase is now albite, and many appear to have been olivine-free. Thin sections show that much fresh clinopyroxene remains and textures are often variolitic. The pillows are commonly highly vesicular. Chert, jasper and sometimes carbonate minerals may occupy spaces between pillows. Sometimes, carbonation (especially dolomitisation) of the pillows has occurred, and whole pillows have been replaced by dolomite. In addition to replacement carbonate, true limestones and dolomites occur. They are brown- and pink-weathering, clastic, laminated rocks which are often cherty and, in rare instances, oolitic. The rose-pink carbonate rocks contain rhodocrosite. The quartzites are slightly feldspathic, pink- and white-weathering rocks, usually medium- to coarse-grained but sometimes conglomeratic. They range from thinly bedded and laminated units to massive units up to 10 m in thickness. They are closely associated in the field with the limestones and dolomites. The greywackes, which range from a few centimetres to a metre in thickness, are often graded and conglomeratic, and the sand-grade fraction includes both volcanic and plutonic fragments. Clasts of a volcanic provenance include andesite, rhyolite and rhyolitic tuffs, whilst clasts of plutonic provenance include granite, acid gneiss and rare schist. The matrix has recrystallised to an aggregate of quartz, muscovite and chlorite. The greywackes commonly occur in association with more pelitic rocks which include thin blue-grey, green and red slates as well as, more commonly, thicker units of similarly coloured, laminated silty pelites and siltstones. Well bedded and laminated, green and blue, basic tuffs (which have been heavily altered to chlorite, epidote and carbonate) are associated with the pillow lavas. Hyaloclastites are common; their originally glassy basic fragments are now represented by aggregates of chlorite, epidote and calcite, but the original textures are often readily seen in thin section. Cherts and jaspers are usually closely associated with the pillow lavas and the carbonate rocks. Those occurring in the inter-pillow spaces are brightest in colour; those which replace pillows and carbonate rocks are more dull. Dark grey, thinly bedded cherts occur within some of the basic tuffs. Matley (1928) noticed that jaspers and cherts occasionally contain abundant spherical microscopic structures resembling *Radiolaria*.

The acid tuffs of the Gwyddel Beds are white- to cream-weathering, but are grey or greenish grey on a fresh surface. They are thinly bedded rather than laminated (2–5 cm thick on average) and often porcellanous. Thin sections show that they are very fine-grained and consist of broken albites in a matrix of now devitrified former glass dust.

Both groups have been metamorphosed and the metamorphic grade increases eastwards. Thus in the west (near Bardsey Sound, for example), the rocks are only weakly metamorphosed but strongly deformed, whereas towards the east they pass into chlorite- and mica-schists and finally into gneisses. The passage eastwards from the low-grade Gwna Group rocks into schists is gradual, as both Matley (1928) and Shackleton (1956) have claimed, and is probably best seen in the coastal exposures north of Trwyn Bychestyn. As the metamorphic grade increases and the rocks become phyllitic, it is still possible to distinguish individual pillows and jasper masses in the basic phyllites, and quartzite and grit bands in the pelitic phyllites. These relict features become less readily distinguishable towards the east, and within 500 m the rocks are thorough-going basic, pelitic and semi-pelitic schists. Further passage eastward from schists into gneisses was claimed by Shackleton (1956), although Matley (1928) thought that the gneisses represented an earlier basement series. Schists can be seen to be followed eastward by migmatitic acid and basic gneisses in a river section southeast of Llangwnadl and in a coastal section at Penrhyn Nefyn, but in each instance the critical junction is probably faulted.

The schists are pale green, well foliated, medium-grained rocks. They are pelitic, semi-pelitic and basic, but commonly consist of lepidoblastic muscovite–chlorite-rich bands alternating with thin granoblastic, psammitic bands and augen-like segregations, both of which now consist of albite and quartz. The chlorite is dirty in thin section and is probably retrogressive after biotite. Garnet is sometimes present. The gneisses range from ultra-mafic to felsic. Ultra-mafic gneisses are coarse-grained, unfoliated, garnet-bearing amphibolites. They may be migmatitic, in which case they carry coarse-grained, feldspathic segregation veins. Mafic gneisses, which may also be migmatitic, are commonly typical amphibolites and consists of nematoblastic amphibole and plagioclase. Felsic gneisses are commonly migmatitic and may carry garnet. The increase in the metamorphic grade from phyllites through schists to gneisses takes place over unusually short distances of perhaps a kilometre or two, and is as rapid as it is in the Penmynydd Zone of Anglesey. Greenly (1919) and Matley (1928) consequently referred to the Llŷn schists as being of the same Penmynydd Zone of metamorphism.

Associated with the gneisses is the unfoliated Sarn Adamellite, which was once regarded by Matley as belonging to the Lower Palaeozoic because it was interpreted as intrusive into Ordovician sediments, but was subsequently shown by Matley and Smith (1936) to be Precambrian. Shackleton (1956) further showed that similar adamellites occur within the gneisses near Meillionydd and elsewhere, and has convincingly demonstrated that the Sarn Adamellite is merely a largely unfoliated member of the gneiss group.

Inland exposures of the complex are poor due to a thick drift cover, so consequently little is known of the structure. All that can be said at present is that the complex shows evidence of several phases of Precambrian folding and that the main fold axes trend northeast–southwest.

Relationships to younger rocks are commonly obscured by drift, but at several localities Arenig sandstones and conglomerates can be seen resting on various formations of the Mona Complex. Matley (1928) believed that the Ordovician junction was faulted, with the Mona Complex thrust southeastward over the Ordovician sediments along a 'boundary thrust'. Discoveries of unconformable relationships (Matley 1932, Matley and Smith 1936) eventually led to the abandonment of the boundary thrust hypothesis. Shackleton (1956) stressed the relationship between the Sarn Adamellite and the gneisses, and showed that a thrust boundary was most unlikely and that the junction was better interpreted as a simple unconformity.

Shackleton (1969) regarded the Mona Complex of Anglesey as essentially eugeosynclinal and cited as convincing evidence the co-existence of pillow lavas, banded jaspery cherts and greywackes. He pointed out, however, that towards the top of the sequence the varied sediments—which include oolitic limestones and ortho-quartzites along with possible ignimbrites—indicate a progressive shallowing and possibly the emergence and establishment of subaerial conditions from time to time. It could be suggested that typically geosynclinal sediments and volcanics were first becoming mixed with, and then succeeded by, miogeosynclinal sediments.

The Mona Complex can be interpreted in terms of plate tectonic theory. The existence of glaucophane schists has long been known from the Anglesey Precambrian and the demonstration by Greenly (1919) that strongly contrasting schists and gneisses carrying sillimanite also occur 7 km away across the strike, allows one to suggest that on Anglesey, and to a lesser extent on Llŷn, we are dealing with the remnants of a paired metamorphic belt which perhaps had affinities with the Mesozoic and Tertiary paired metamorphic belts of the circum-Pacific region (Miyashiro 1961). If this is so, then we may go on to suggest that the pillow lavas, basic tuffs, cherts and jaspers of the Gwna Group accumulated on an ocean floor, and that ocean floor spreading caused them to slough off in a trough at a destructive plate margin where greywackes, pelitic rocks, etc were added, so constituting the mélange. The sequence on Anglesey was subsequently metamorphosed to 'glaucophane schist facies' in the subduction zone, whereas to the continental side of the plate margin more variable shelf sediments and great volumes of acid volcanics accumulated and eventually underwent andalusite–sillimanite-type regional metamorphism.

Such an interpretation can be criticised in detail. For example, in the circum-Pacific belt, the type locality for paired metamorphic belts, rocks of the jadeite–glaucophane type normally occur on the oceanward side of the belts. Plate tectonic interpretations of the Mona Complex (see, for example, Dewey 1969a) have illustrated the subduction zone as dipping towards the southeast beneath a continental plate. This implies that the jadeite–glaucophane-type metamorphic rocks should occur to the northwest of the andalusite–sillimanite-type schists and gneisses. In fact, this simple pattern does not hold on Anglesey because granitic gneisses and migmatites are well developed in the north and northeast, whereas lawsonite–glaucophane rocks are best developed in the southern half of the island. If we wish to retain this interpretation it becomes necessary to resort to special pleading.

It could be suggested that the original metamorphism of the Mona Complex encompassed a range from the lawsonite–albite facies to the greenschist facies, and that the formation of high-grade gneisses and migmatites was a slightly later phenomenon directly related to the emplacement of granites such as the Coedana Granite of Anglesey and the Sarn Granite of Llŷn. This could be supported by the suggestion of Oxburgh and Turcotte (1971) that the continental plate grows oceanward by the accretion of oceanic crust and sediment as long as subduction continues. This means that the subduction zone is forced to migrate oceanward, as also must the zone of magma generation. In turn, the region of high-temperature metamorphism migrates similarly and eventually may come to impinge upon the somewhat earlier formed glaucophanitic metamorphites.

Although this explains the disposition of different types of metamorphites in mid- and southeastern Anglesey and the disposition of metamorphic grade in Llŷn, it fails to explain the high-grade migmatitic gneisses of northern Anglesey, and it also fails to take account of the evidence derived from the Lower Cambrian of Gwynedd which, as we shall see, indicates that much of the clastic detritus was derived from a segment of plutonic continental crust away to the northwest. It is still possible, however, to accommodate the evidence within a plate tectonic framework if it is suggested that the gneisses of Anglesey are of two different ages: the gneisses of the north may represent part of a raft of pre-Mona Complex continental crust carried on the essentially oceanic plate, and that subduction ceased when all the oceanic crust in advance of the raft had been consumed. This suggestion might also explain why it is that the structures in the Mona Complex verge consistently southeastward.

It may therefore be that two groups of high-grade metamorphites are present on Anglesey: those associated with the granitic intrusive rocks and the Penmynydd Zone of metamorphism; and an earlier suite representing pre-existing basement. In the light of these difficulties it is worth reiterating that Greenly (1919) and Matley (1928) believed that all the gneisses were older than the bedded, low-grade rocks and the schists,

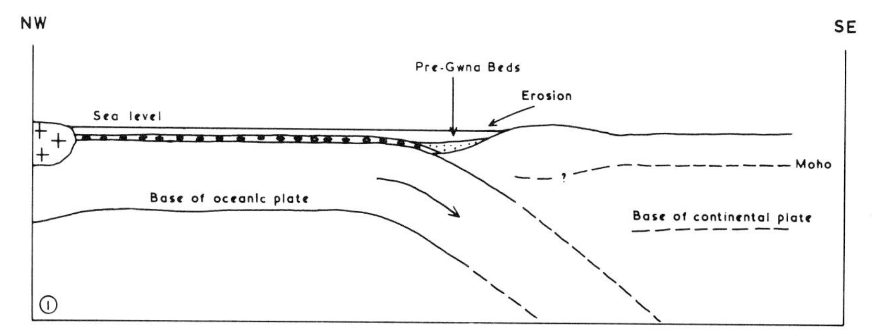

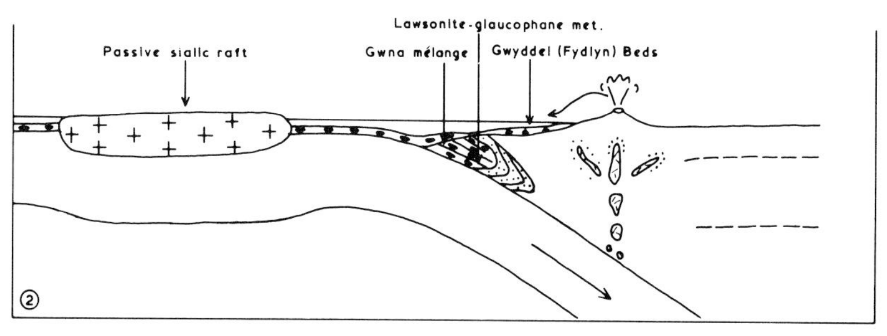

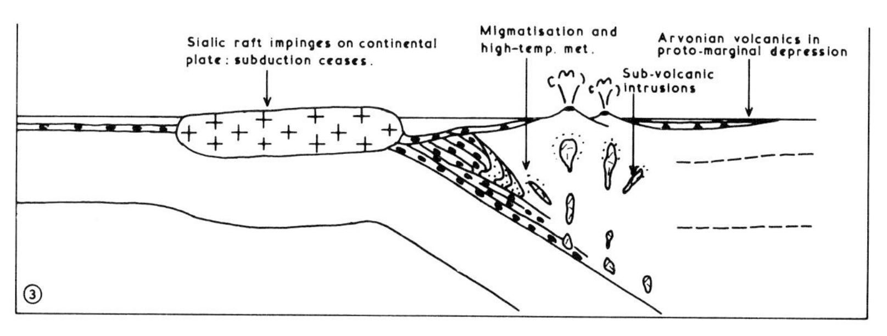

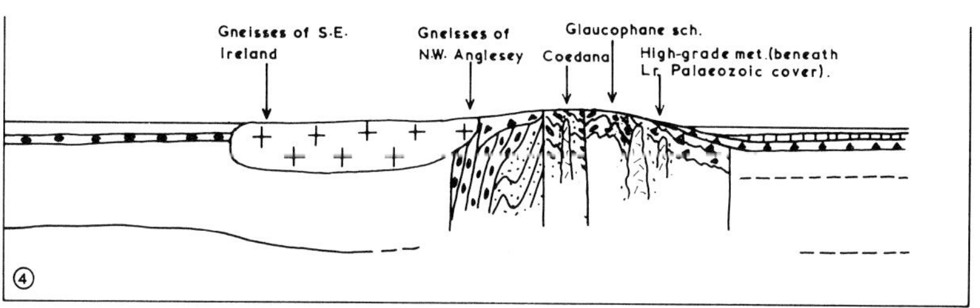

Figure 2. An illustration of a plate tectonic interpretation of the formation of the Mona Complex, Arvonian and Cambrian rocks. (1) Start of subduction: accumulation of pre-Gwna beds in trough. Eugeosynclinal sequence of greywackes, shale, etc. (2) Continued subduction: sediments and ocean floor mafics carried down, resulting in lawsonite–glaucophane metamorphism. Gwna mélange accumulates. Acid volcanism produces the Gwyddel (Fydlyn) Tuffs. (3) Maximum subduction: subduction zone and volcanic front have migrated oceanward due to accretion. Granites invade the lawsonite–glaucophane metamorphites. Acid volcanism leads to Arvonian ignimbrites in a marginal depression. Subduction ceases when the sialic raft impinges upon the edge of the continental plate. (4) Post subduction: uplift due to buoyancy leads to the 'Irish Sea landmass'. The marginal depression is invaded by the sea and marine Cambrian sediments accumulate.

whereas Shackleton (1956) took the view that they were all essentially of the same age.

The metamorphic rocks of western Llŷn are overlain unconformably by essentially unmetamorphosed Ordovician (Arenig) rocks, so that prior to the Ordovician, the Mona Complex suffered uplift and deep erosion. Part of this large gap in the geological history of the area is filled by another group of Precambrian rocks, the Arvonian Volcanic Group, which occurs in Arfon in the north of the area.

2. The Eocambrian Arvonian Series of the Bangor and Padarn Ridges

Eocambrian rocks are exposed in two ridges which trend northeast–southwest firstly from Bangor to Caernarfon (the Bangor Ridge), and secondly from Bethesda to beyond Pen-y-groes (the Padarn Ridge). Together, the rocks constitute the Arvonian Series, so named by Greenly (1930). The rocks have been described by Greenly (1945) and Wood (1969), although earlier descriptions, such as those of Bonney (1879) and Wynne-Hughes (1917), exist. The rocks consist mainly of rhyolitic volcanics, particularly ignimbrites, but lesser volumes of ash-fall tuffs, agglomerates, intrusive rhyolites and granites are also present. The predominance of welded, eutaxitic, rhyolitic ignimbrites indicates that the bulk of the sequence accumulated subaerially. It is interesting to note that Hughes (1917) came very close to recognising ignimbrites for what they are now thought to be, and his Plate 1 illustrates three beautifully welded tuffs (photomicrographs 1, 3 and 6).

The base of the Arvonian succession is not seen, but nevertheless the volcanic rocks are more than 1000 m thick, and it is clear that many ignimbrite flows are present. Eutaxitic textures are very well preserved and, surprisingly, are much more readily seen in thin section than they are in the younger, better known Ordovician ignimbrites. The ignimbrites of the Padarn Ridge, which occur in the core of a major periclinal anticline, are conformably overlain by a conglomerate which, in turn, is succeeded by greywackes, further conglomerates, thin slates and volcanic horizons (which include more ignimbrites), and then by the undisputed Lower Cambrian succession of slates with minor greywackes. Wood (1969), in a detailed description of the passage from Arvonian into Cambrian strata at Llyn Padarn, suggested that the base of the Cambrian at this locality be placed at the base of the first conglomerate overlying the acid Arvonian volcanics. This means that succeeding sedimentary *and pyroclastic* rocks are to be regarded as Cambrian.

In the Bangor Ridge the situation is different. The undoubted Arvonian rocks are similar to those of the Padarn Ridge and they include the essentially sub-volcanic Twt Hill Granite which grades from a coarse, two-feldspar granite at Caernarfon through a microgranite to a rhyolite at Port Dinorwic. In the Bangor area itself, however, Greenly and Wood agree that the Arvonian tuffs and agglomerates are overlain unconformably by conglomerates which are referred to the Cambrian and which, in their turn, are overstepped by the Ordovician so that the Ordovician comes to rest on the Arvonian. There is thus conformable passage from Arvonian into Cambrian in the southeast at Padarn, but a small break with disconformity at the base of the Cambrian at Bangor. Greenly, however, included in the Arvonian a series of fine-grained, tuffaceous sediments and tuffs from the Bangor area, but since these overlie the lowest conglomerate they are more usefully considered, as Wood has suggested, as part of the Cambrian rather than the Arvonian succession.

It will be recalled that in western Llŷn, and indeed at most localities on Anglesey too, the topmost beds of the Mona Complex are overlain unconformably by Ordovician strata and that Cambrian rocks, with one possible exception, namely a small area of conglomerate at Trefdraeth on Anglesey, are absent. The Trefdraeth conglomerate rests unconformably upon rocks of the Gwna Group and is lithologically unlike any Ordovician conglomerate. It is accepted as Cambrian by both Greenly and by Wood (1969). Again on Anglesey (but not in Llŷn) an ignimbrite, the Bwlch Gwyn Felsite near the Holland Arms, together with tuffaceous sediments at Beaumaris, also rest unconformably upon rocks of the Mona Complex, and are thought to be Arvonian in age. If the ages of the Trefdraeth conglomerates and the Holland Arms ignimbrites are accepted, then both Arvonian and Cambrian rest unconformably upon the Mona Complex in Anglesey,but the base of the Cambrian oversteps the Arvonian onto the Mona Complex towards the northwest.

The situation is represented diagrammatically in figure 3.

In western Llŷn, where Arenig rocks rest unconformably upon Mona Complex metamorphites, it becomes clear that the basal Arenig unconformity spans a time interval represented by the Arvonian plus the

Cambrian. At Padarn it may well be, as Shackleton suggested, that there is passage from concealed Mona Complex through Arvonian into Cambrian, but this suggestion raises problems concerning the timing of the deformation and metamorphism of the Mona Complex. These problems will be considered later (pp 29–30).

3. The Relationship of the Arvonian to the Mona Complex in Terms of Plate Tectonic Theory

The evidence of Trefdraeth and Holland Arms in southeastern Anglesey suggests that the Mona Complex was deformed and metamorphosed prior to the accumulation of the Arvonian. It is likely, as has been suggested, that the sediments and igneous rocks accumulated at a destructive plate margin, and it is possible that the tectonism and related metamorphism resulted in a paired metamorphic belt consisting of rocks of the lawsonite–glaucophane facies of Winkler (1967), together with greenschist and amphibolite facies rocks. The plate margin presumably trended northeast–southwest with the subduction zone dipping to the southeast beneath a continental lithospheric plate.

The presence of the mélange in both Llŷn and Anglesey and of serpentinites in Anglesey is further evidence in favour of the existence of a former subduction zone at a convergent plate margin. The mélange may partly represent sediments and ophiolites which were scraped from the descending oceanic plate and were plastered to the inner wall of the trench.

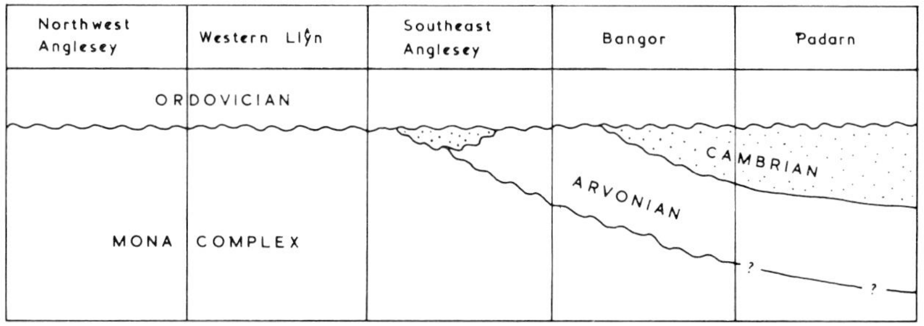

Figure 3. Diagrammatic representation of the stratigraphical relations between Ordovician, Cambrian, Arvonian and Mona Complex rocks.

The Arvonian ignimbrites and other volcanic rocks indicate intense, subaerial, acid volcanism to the southeast of the subduction zone on the continental lithospheric plate. The Arvonian is therefore intimately related by this hypothesis to the earlier metamorphites of the Mona Complex, as are the 'granites' of Coedana on Anglesey and of Sarn in western Llŷn. The relationship is illustrated in figure 2.

CAMBRIAN

Rocks of known Cambrian age crop out in five areas: St Tudwal's peninsula in western Llŷn; the Cambrian Slate Belt which stretches northeast from Nantlle through Llanberis to Bethesda and beyond; northwest of the Padarn Ridge; in the Ynyscynhaiarn Anticline; and in a series of small inliers west and northwest of Moel Hebog. Figure 4 is a correlation chart of the beds exposed in the five areas.

The rocks of Ynyscynhaiarn, St Tudwal's, and the Moel Hebog area can be related to each other without too much difficulty, but the rocks of the Slate Belt are unique, so that correlation with the first group is difficult and is still in dispute.

1. The Cambrian of the Slate Belt

It will be recalled that the Cambrian succession is taken to begin with the first conglomerate at the top of the Arvonian volcanic succession at Llyn Padarn (Llanberis) and Bangor. At Padarn, the conglomerate is succeeded immediately by a rather varied succession of ignimbrites, air-fall tuffs and agglomerates, conglomerates, sandstones and slates. At Bangor, the basal conglomerate is disconformable upon the Arvonian volcanic rocks. The interpretation of Greenly (1945, 1946), that the basal Cambrian is in turn overlain by a thrust sheet of isoclinally folded

Figure 4. Cambrian correlation chart. (In large part after Cowie *et al* 1972, with permission.)

	Series	Zone		1 Nantlle	2 Llanberis–Bethesda
UPPER	Tremadoc	*Angelina sedgwicki*		—	—
		Shumardia pusilla		—	—
		(No fossils yet of zonal significance)		—	—
		Clonograptus tennellus		—	—
		Dictyonema flabelliforme		—	—
	Merioneth	*Acerocare*		—	—
		Peltura, etc		—	—
		Leptoplastus		—	—
		Parabolina spinulosa		—	Carnedd y Filiast Grit and Ffestiniog Beds 300 m
		(No fossils yet of zonal significance)		Ffestiniog Beds 426 m	
		Olenus		Maentwrog Beds 346 m	Maentwrog Beds 300 m
		Agnostus pisiformis		—	—
MIDDLE	St David's	*Paradoxides forchhammeri*		—	—
		Paradoxides paradoxissimus	*Pt. punctuosus*	—	—
			H. parvifrons	—	—
			Pt. atavus, *T. fissus*	—	—
			Pt. gibbus	—	—
		Paradoxides oelandicus		Cymffyrch Grit 200 m	Bronllwyd Grit 350 m
LOWER	Comley	Protolenid–Strenuellid		Green Slate Mottled Blue Slate Pen-y-bryn Grit Striped Blue Slate Dorothea Grit Purple Slate } 745 m	Green Slate Victoria Red and Grey Slates Dwndwr Grit Striped Blue, Purple and Red Slates Lower Grit Group Lower Red and Blue Slates } 1800 m
		Olenellid			
		Non-trilobite zone		Glôg Grit Group 600 m Cilgwyn Conglomerate 400 m Tryfan Grit Group 300 m	Conglomerate–Grit–Tuff Group 450 m
				Arvonian Volcanic Series	Arvonian Volcanic Series

— Firm correlation with the standard; - - - possible error of ±1 zone; · · · · · considerable uncertainty.

1. Cattermole and Jones (1970), Morris and Fearnsides (1926); 2. Williams (1927), Williams (1930), Wood (1969); 3. Ramsay (1866); 4. Nicholas (1915, 1916), Crimes (1970a,b). 5. Fearnsides (1910). 6. Shackleton (1959).

3 Northwest of Padarn Ridge	4 St Tudwal's	5 Ynyscynhaiarn	6 N and W of Moel Hebog
Slate Group	Tremadoc Beds 80 m (faulted boundaries)	TREMADOC SLATES: Garth Hill Beds 36 m	Dolgelly Beds 6 m
Conglomerate–Grit–Tuff Group	Ffestiniog Beds	Penmorfa Beds 30 m	Ffestiniog Beds 300 m
Arvonian Volcanic Series	Maentwrog Beds 200 m	Portmadoc Flags 60 m	? ? ? Base not seen
	Nant Pîg Mudstones	Moelygest Beds 70 m	
	Upper Caered Mudstones 90 m	Unnamed Slates	
	Caered Flags 30 m	Dictyonema Band & Tynllan Beds 65 m	
	Lower Caered Mudstones 33 m	LINGULA FLAGS: Dolgelly Beds 70 m	
	Cilan Grits 300 m	Ffestiniog Beds 300 m	
	Manganese Beds 140 m	Maentwrog Beds	
	Hell's Mouth Grits 170 m	? ? ? Base not seen	
	? ? ? Base not seen		

Arvonian tuffs, was rejected by Wood (1969). Instead, as at Padarn, the basal conglomerate was shown to be overlain by further tuffs and sandstones which, in turn, are overstepped by Arenig sediments. Around Nantlle, Morris and Fearnsides (1926) described the Arvonian rhyolites (actually ignimbrites) comprising the Clogwyn Volcanic Group as being overlain by the Tryfan Grit Group. They took the succeeding Cilgwyn Conglomerate to be the base of the Cambrian and thought that it rested discordantly on the underlying Tryfan Grit and Clogwyn Volcanic Groups. Cattermole and Jones (1970) continued to regard the Clogwyn Volcanic Group as Arvonian, but preferred to consider that the Tryfan Grit Group represented the lowest Cambrian sediments, and that the Tryfan Grit Group rested more or less conformably upon the Arvonian. Crimes (1970a) went even further and regarded the Arvonian and all succeeding clastic sediments as Cambrian. Whatever preference is accepted, however, the coarse-grained Cilgwyn Conglomerate passes up into the Glôg Grit and thence into a series of greywackes, siltstones and slates, which all workers regard as Cambrian.

The Tryfan Grit Group is about 300 m thick and consists of a lower division of coarse-grained, feldspathic sandstones showing large-scale tabular cross bedding, and an upper division of finely parallel-laminated, white-weathering siltstones. The siltstones are dark grey to green on freshly fractured surfaces, and may represent re-worked vitric tuffs. The succeeding Cilgwyn Conglomerate is up to 400 m thick and contains well rounded pebbles of Arvonian volcanics up to 15 cm in diameter, as well as pebbles of jasper and quartzite which can be matched with the jaspers and quartzites of the Gwna Group of the Mona Complex. The pebbles have been strongly flattened in the plane of the cleavage. The Glôg Grit comprises a group of coarse- and fine-grained sandstones and quartzites with subordinate thin slaty beds. Large-scale tabular cross bedding is common. The evidence points to a shallow-water environment of deposition for much of the Tryfan Grit, Cilgwyn Conglomerate and Glôg Grit groups.

The Slate Group which follows is about 750 m in thickness. It consists largely of green, purple and blue slates, with subordinate thin siltstones and two greywacke horizons. The group crops out to form the 'Cambrian Slate Belt' from which some of the world's finest roofing slates have been obtained. The two greywacke horizons, the Dorothea and Pen-y-bryn Grits, each consist of a series of greywacke beds from 50 cm to 2 m thick with intercalated slates. Individual greywacke units commonly show a graded interval passing up into an interval of parallel lamination which may be convolute, and rarely the unit is completed by an interval of cross lamination. The sediments probably represent proximal turbidites. Sole markings include flutes and grooves. Mud pellets are common towards the base of graded intervals. Sand- and granule-grade clasts include abundant vitreous blue quartz grains which cannot readily be matched with Arvonian or Mona Complex quartz. Other clasts include occasional jasper and greenschist, presumably derived from the Mona Complex. The siltstones usually show parallel lamination, which may be convoluted in some cases. Cross lamination is common, and some laminae show grading. The evidence suggests that the slates reflect a deepening of the basin; greywackes represent proximal turbidites, whereas the siltstones may represent distal turbidites. The rocks of the Slate Belt are strongly deformed and dips are usually very steep so that the few directional sedimentary structures cannot be used to determine satisfactorily palaeocurrent directions. A sparse Protolenid fauna with *Pseudatops* (*Conocoryphe*) *viola* (Woodward) has been obtained from green slates at the top of the succession, thus indicating a late Lower Cambrian age according to Howell and Stubblefield (1950).

The rocks of the slate group are followed abruptly by the 200 m thick Cymffyrch Grit and its lateral equivalents, that is, the Bronllwyd and Dinas Grits. The formation consists of a series of greywackes and conglomeratic greywackes each up to 6 m thick, but commonly 50 cm thick, with subordinate, thin interbedded slates and siltstones. In the lower part of the succession the conglomeratic greywackes often carry large mud pellets. Most greywacke beds consist of a graded interval followed by an interval of parallel lamination, although a large minority consist of just the interval of parallel lamination which is sometimes convolute. The cross lamination interval does occur, but this is uncommon. Sole markings include flutes, grooves and prod marks. The evidence suggests the rocks are proximal turbidites. Sand- and granule-sized clasts include material of both plutonic and volcanic provenance: gneiss, granite and schist on the one hand, and rhyolite, trachyte and andesite on the other.

Morris and Fearnsides (1926) believed the lower boundary of the Cymffyrch Grit to be faulted, but Cattermole and Jones (1970) claimed that in at least two localities there is passage from the topmost slates into the Cymffyrch Grit. They therefore regard the Cymffyrch Grit and its lateral equivalents as topmost Lower Cambrian, as also does Crimes (1970a). In the Vale of Nantlle, the Cymffyrch Grit is overlain by Llanvirn slates, the junction having been interpreted by Morris and Fearnsides (1926) and Roberts (1967) as a fault. To the northeast, near Mynydd Mawr, the Cymffyrch Grit is faulted out and rocks ascribed to the Maentwrog Beds lie against the Cambrian Slates. The Maentwrog Beds consist of about 350 m of black silty slates, grey banded siltstones with thin sandstones, and a prominent pisolitic ironstone some 10 m thick near the middle of the succession. The siltstones show both parallel lamination and ripple-drift bedding. The pisoliths in the ironstone consist of a chamosite-like mineral and the siltstones immediately beneath the horizon are burrowed. Thus, although much of the Maentwrog sequence may have been laid down in deeper water, a period of shoaling is indicated by the ironstone. The Maentwrog Beds pass upward into the Ffestiniog Beds which comprise slates, siltstones, sandstones and quartzites, and together these reach about 450 m in thickness.

In the northeast of the Slate Belt, coarse sandstones, granule conglomerates and greywackes constitute much of the Ffestiniog succession. The beds are generally well sorted and some tend to wedge out laterally in a few metres. Large-scale tabular cross bedding is common, with the lower bounding surface to sets being sometimes strongly erosional. Upper surfaces are commonly ripple-marked and the forms include symmetrical, asymmetrical, flat-topped, linguoid and cuspate types. Wave lengths reach up to 1 m but are commonly less; amplitudes range up to 10 cm. *Lingulella* occurs in the interbedded mudstones and siltstones. Trace fossils include abundant *Cruziana*; and *Rusophycus* and *Phycodes* are common. Crimes (1970a,b) has described *Diplichnites*, *Dimorphichnus* and *Planolites*. Towards the top of the succession *Skolithos* is found. The sedimentary and biogenic structures suggest an accumulation in shallow water, probably between low water mark and wave base.

The Cambrian succession is terminated either by faulting or by a sub-Ordovician unconformity.

2. The Cambrian Northwest of the Padarn Ridge

This area of Arfon is poorly exposed and details of its geology are little known. Ramsay (1866) showed that the succession consists of greywackes, tuffs and conglomerates overlain by coloured slates comparable with those of the Nantlle–Llanberis slate belt. The thicknesses of the beds are unknown. Greenly (1944, 1945) however, believed that the tuffs, tuffaceous sediments and sandstones overlying conglomerates at Bangor were Arvonian rather than Cambrian. Wood (1969) rejected Greenly's views, and it is Wood's view which is accepted here.

3. The Cambrian of St Tudwal's Peninsula

The Cambrian rocks of the restricted outcrop in St Tudwal's peninsula were described by Nicholas (1915, 1916) and, in part, more recently by Bassett and Walton (1959) and Crimes (1970a). The rocks are for the most part lithologically unlike those of the Slate Belt and, instead, have strong similarities to those of the Harlech Dome some 30 km away to the east. Over 700 m of coarse-grained greywackes and slates make up most of the succession and the rocks have yielded a sparse Protolenid fauna showing them to be late Lower Cambrian in age. The remainder of the Cambrian succession is much thinner, with representatives of both the Middle and Upper Cambrian present.

The succession begins with the Hell's Mouth Grits, a formation consisting of grey, green and blue greywackes with interbedded siltstones and mudstones. The greywackes are commonly graded and show a variety of sole markings, including flutes, grooves, prod marks, etc. A variety of ripple marks have also been described. Feeding burrows have been described by Crimes (1970a). Currents appear to have flowed from the northeast and the detritus has been derived from rocks similar to those of the Mona Complex. The greywacke/shale formation is overlain by the Mulfran, or Manganese Beds which consist of manganese-rich shales, mudstones and greywackes. The mudstones are well laminated with ripple-drift bedding; the greywackes are about 0·5 m thick and are commonly graded. The succeeding Cilan Grits consist of graded greywackes with thin mudstones. Mud-pellet greywackes are present, as are various sole structures. Washouts and both tabular and trough cross bedding are present. Apart from the fossils in the Hell's Mouth Grits and various trace fossils, the succession up to this horizon is largely unfossiliferous and probably all belongs to the Lower Cambrian. The succeeding Caered Flags and Mudstones begin with a porcellanous, vitric tuff at the base which gives way to a greywacke sequence with beds up to 1 m thick. Siltstones, with convolute lamination and mudstones, often ripple-marked, alternate with greywackes. The siltstones may show sole markings and ripple-drift lamination. The upper part of the formation is predominantly mudstone and siltstone and has yielded a Middle Cambrian fauna. The overlying black, pyritous Nant Pîg Mudstones are fossiliferous: they have yielded an abundant trilobite fauna of Middle Cambrian age and, in places, are strongly bioturbated.

The succeeding Maentwrog Beds begin with a calcareous conglomerate which is followed by greywackes and shales, siltstones and mudstones, and then by shales with just occasional sandstones. A great variety of flute casts, groove casts and sole markings may be seen, as well as load casts and flame structures. Convolute lamination is well developed in the siltstones and the tops of many beds are ripple-marked. Representatives of the Ffestiniog Beds occur on St Tudwal's Island and consist of shales, siltstones and fine-grained greywackes. Towards the top of the successon, an horizon very rich in *Lingulella* occurs constituting the 'Lingulella Band'. The formation is rich in trace fossils including *Cruziana* and *Skolithos*.

4. The Cambrian of the Ynyscynhaiarn Anticline

The sequence includes members of the Lingula Flags succeeded by the Tremadoc Series, and was described by Fearnsides (1910) and in part by Shackleton (1959). The succession occupies a northerly plunging anticline and begins with about 100 m of slates and greywackes assigned to the Maentwrog Beds. The sediments closely resemble the Maentwrog Beds of St Tudwal's. The succeeding Ffestiniog Beds consist of 600 m of interbedded greywackes and subordinate shales which, again, closely resemble the Ffestiniog Beds of St Tudwal's. The Dolgelly Beds are 70 m thick and consist essentially of black, pyritous slates and laminated mudstone with just a few thin, fine-grained greywackes and siltstones. The siltstones are sometimes calcareous and may show ripple-drift lamination. A thin crystal tuff, 10–30 cm thick occurs near the middle of the succession. The Lingula Flags contain a good Olenid fauna and are also rich in *Lingulella*.

The Tremadoc Beds are conveniently grouped with the Cambrian, despite some faunal affinities with the Ordovician, because the Tremadoc Beds lie conformably upon the Lingula Flags, whereas the basal Arenig rests unconformably upon older rocks. The Tremadoc sediments are essentially slates with siltstone and thin, graded greywacke horizons. The succession begins with 65 m of grey slates and parallel-laminated siltstones (termed the *Niobe* Beds or Tyn-llan Beds) which, at certain horizons, yield abundant trilobites. They are succeeded by the poorly exposed, 6 m thick *Dictyonema* band. This is a parallel-laminated slate or sometimes silty slate which, in addition to containing abundant *Dictyonema*, also carries trilobites and sponge spicules. The succeeding Moel-y-gest Beds are 70 m thick and consist of sparsely fossiliferous grey slates and mudstones with occasional thin, silty laminae and rare, graded, fine-grained greywackes. The proportion of sand and silt laminae increases upward so that there is a transition towards the top into the overlying Portmadoc Flags. These consist of parallel-laminated silts with very thin greywackes, and the formation is over 60 m thick. The Penmorfa Beds are mainly blue-black pyritous slates with a few silty and sandy laminae. They are over 30 m thick and certain horizons contain abundant, usually, distorted, trilobites. The Tremadoc succession closes with the 35 m thick Garth Hill Beds, which consist of slates, silts and fine greywackes.

5. Cambrian Inliers West and North of Moel Hebog

Shackleton (1959) recognised Cambrian rocks in a series of essentially periclinal inliers in Cwm Pennant and to the north of Moel Hebog. Representatives of the Ffestiniog Beds occur in each inlier and in one they are accompanied by a few metres of Dolgelly Beds. In the southernmost inlier, the Ffestiniog Beds are about 200 m thick and closely resemble the Ffestiniog Beds of the Tremadog area. They consist of green and bluish grey laminated silty slates with thin greywackes. Cross lamination is common. In the northern inliers three massive quartzites appear and the greywackes become less important. About 330 m are seen. The Dolgelly Beds are seen in only one locality in the north. They consist of 6 m of dark blue-black, finely laminated, silty slates and have yielded a diagnostic trilobite–brachiopod fauna.

6. Deposition of the Cambrian

We have seen that the metamorphism and deformation which affected the Precambrian Mona Complex was accompanied and thereafter followed by acid volcanism which, especially away to the southeast, gave rise to the Arvonian Series. In terms of plate tectonic theory, we may assume that the subduction zone became inactive and an epicontinental sea developed to the southeast as the shallow subsiding trough in which the Arvonian had accumulated continued to subside. Marine Cambrian sediments thus begin with conglomerate which, in places, succeeds the Arvonian conformably. Elsewhere, as the sea broadened and deepened, conglomerates and sandstones rest unconformably upon Arvonian and even, as at Trefdraeth on Anglesey, upon Mona Complex. Perhaps plate tectonic theory would suggest a comparison with the present-day Japan Sea?

Crimes (1970a) has published a facies analysis of the Cambrian rocks of Wales. He showed that in Snowdonia and Llŷn marine deposition began with a shallow-water clastic sequence, but rapid deepening of the sea led to a change to a sequence of interbedded mudstones with turbidite greywackes and siltstones. Some of the early greywackes were derived from an Irish Sea landmass which was close by to the northwest and which, we may assume, had arisen (in terms of plate tectonics) on the site of the now dead subduction zone. The borders of the trough were aligned northeast to southwest, the grain having been inherited from the Precambrian structures. In the northwest was the Irish Sea landmass, whilst to the southeast was the English Midlands and Welsh Borderlands landmass. The supply of clastic material was reduced towards the close of the Lower Cambrian, so allowing the deposition of manganese-rich sediments in quiet water. Uplift of the southeastern border led to an influx of detritus near the base of the Middle Cambrian but, with time, the supply was reduced and higher Middle Cambrian sediments are finer-grained and were derived from currents flowing from the south and southwest. A hiatus occurs at the top of the Middle Cambrian. This was followed in the Maentwrog Stage by turbidity currents bringing in detritus from a westerly direction and, as this supply waned, so finer material was brought in from the south. The Ffestiniog Stage saw fine material still derived from the south, but coarse-grained material in Arfon indicates close proximity to a shoreline. The Dolgelly and Tremadoc Stages were periods of quiet, shallow-water accumulation of largely fine-grained sediments.

Following the Tremadoc, the sediments were folded and a period of uplift and erosion ensued, the uplift and erosion being strongest in the west. Sedimentation recommenced in the Arenig Stage of the Ordovician.

Finally, it may be worth drawing attention to the postulated Irish Sea landmass. Volcanic detritus is not a dominant component of the Cambrian sediments; indeed it is restricted to spilitic types along with some rhyolitic and andesitic material. Tuffs are only represented at St Tudwál's in the Caered Flags and in the Dolgelly Beds at Ynyscynhaiarn. Most of the detritus is of plutonic and metamorphic origin. Plate tectonic theory would prefer the landmass to have been an extinct volcanic arc, yet clearly it was a slice of plutonic and metamorphic basement.

ORDOVICIAN

Folding and uplift at the end of the Tremadoc led to erosion and planation. The effects were greatest in the west and northwest, where several thousands of metres of Cambrian strata were stripped off to reveal the underlying Precambrian. Submergence followed with the incursion of the Ordovician sea and re-establishment of marine conditions. Thus on Anglesey and Arfon in the northwest, and in southwestern Llŷn, west of St Tudwal's, basal Arenig sediments rest on various members of the Precambrian.

The Ordovician has been divided into five series on the bases of graptolite and shelly faunas. The Caradoc and Ashgill Series have been further subdivided into eleven stages on the basis of shelly faunas alone, but no stages have yet been defined in the Arenig, Llanvirn and Llandeilo

Figure 5. Ordovician correlation chart. (In part after Williams *et al* 1972, with permission.)

Series	Stages (where defined)	Graptolite zones	1 Aberdaron–Rhiw	2 St Tudwal's	3 Llanbedrog–Pwllheli	4 Llwyd Mawr–Llanystumdwy	5 Southwestern Snowdonia
ASHGILL	Hirnantian						
	Rawtheyan	*D. anceps*			Crugan Mudstone	Grey-green mudstones	
	Cautleyan						
	Pusgillian	*D. complanatus*					
CARADOC	Onnian	*P. linearis*			Nod Glas	Black shales	
	Actonian	*D. clingani*					
	Marshbrookian						
	Longvillian				Penarwel Drive Beds Llanbedrog Volcanic Group Mynytho Volcanic Group Llanbedrog Sandstones and Mudstones	Blue-black mudstones Llanystumdwy Volcanic Group Llwyd Mawr Ignimbrite	Snowdon Volcanic Group: Upper Rhyolitic Group Middle Basic Group Lower Rhyolitic Group
	Soudleyan	*C. wilsoni*					Upper Glanrafon Beds: Gorllwyn Grits Gorllwyn Slates Pren-teg Grits Portreuddyn Slates
	Harnagian	*C. peltifer*					
	Costonian	?		Pen-y-Gaer Mudstone Hen-dy-Capel Ironstone			Y Glog (Moelwyn) Volcanic Gp Lower Glanrafon Beds: Tiddyn-dicwm Beds
LLANDEILO	Upper	*N. gracilis*					
	Middle						
	Lower	*G. teretiusculus*			? Garn Fadrun Shales		
LLANVIRN	Upper	*D. murchisoni*			TAL-Y-FAN ARGILLITES: Grey and black shales	Maesgwm Slates	Maesgwm Slates
	Lower	*D. bifidus*	Shales Ignimbrites Shales with tuffs		Grey and brown shales		
ARENIG	Upper	*D. hirundo*	Pillow lava Black shale		Grey and black shales	Grey-black mudstones, siltstones and slates	Upper Pennant Quartzite Green Slates
	Lower	*D. extensus*	Bodwrdda Slates Porth Meudwy Beds Meudwy Valley Slates Parwyd Grit	Llanengan Mudstones Tudwal Sandstone	Pen Benar Shales Pen Benar Conglomerate	Laminated siltstones Conglomeratic sandstone	Pant-y-wrach Beds Garth Flags Garth Grit

1. Crimes (1969), Matley (1938); 2. Crimes (1969), Nicholas (1915); 3. Crimes (1969), Fitch (1967), Matley (1938); 4. Harper (1956), Roberts (1967); 5. Shackleton (1959); 6. Williams (1927); Williams and Bulman (1931); 7. Evans (1968); 8. Davies (1936), Stevenson (1971); 9. Elles (1909), Stevenson (1971), Wood and Harper (1962); 10. Davies (1936), Howells *et al* (1973).

6 Mid-Snowdonia	7 Northeastern Snowdonia	8 Tal-y-fan	9 Conwy	10 Capel Curig–Trefriw
			Conway Castle Grits Deganwy Mudstones Bodeidda Mudstones	Grinllwm Slates Trefriw Mudstone and Tuff
Black slates		Llanrhychwyn Slates	Cadnant Shales	Llanrhychwyn Slates
Snowdon Volcanic Group: Upper Rhyolitic Group; Bedded Pyroclastic Group; Lower Rhyolitic Group	Bedded Pyroclastic Group Lower Rhyolitic Group	Crafnant Volcanic Group: U: Basic tuff; M: Rhyolitic tuff; L: Basic tuff		Crafnant Volcanic Group: U: Tuffs, tuffaceous sediments and slates; M: Tuffs, tuffaceous sediments and slates; L: Rhyolitic tuffs and slates
Upper Glanrafon Beds: Gwastadnant Grits Lower Glanrafon Beds: Glanrafon Slates	Upper Glanrafon Beds Capel Curig Volcanic Group Middle Glanrafon Beds Carnedd Llywelyn Volcanic Group Lower Glanrafon Beds: Black Slates; Pisolitic Ironstone	Glanrafon Beds: Slates and tuff; Siltstones and slates Conway Volcanic Group: Maen Amor Tuff; Maen Amor Sandstones; Gyrach Andesitic Tuff; Fairy Glen Slates Purple and grey slates and sandstones ? ? ? (Base not seen)	Coetmor Group: siliceous grits Conway Volcanic Group: Coetmor Group Volcanics; Coetmor Ash; Upper Brecciated Lava Group; Bodlondeb Ash Group (Base not seen)	Glanrafon Beds: slates, sandstones, siltstones, mudstones, with thin acidic and basic tuffs Capel Curig Volcanic Group Sandstones (Base not seen)
Maesgwm Slates	Blue-black slates Grey sandstone Blue-black slates Basal grit			

Series. The subdivisions, together with the graptolite zones, as recommended by Special Report No. 3 of the Geological Society of London (1972) are given in the Ordovician correlation chart shown in figure 5.

1. Arenig

Wherever the base of the Arenig is exposed it is seen to be a plane of unconformity, and the basal Arenig sediments overstep northwestwards from the Tremadoc sediments of the Harlech Dome and the Ynyscynhaiarn Anticline across the Lingula Flags, Middle and Lower Cambrian in St Tudwal's, onto Precambrian in western Llŷn and Arfon.

In western Llŷn, Crimes (1970b) distinguished a number of facies on the bases of lithology, inorganic sedimentary structures and biogenic structures. The following descriptions are based on his work. He recognised a Conglomeratic Facies, which is usually the basal member and which contains *Bolopora undosa* (Lewis), formerly thought to be a polyzoan, but which is now regarded by Hofmann (1975) as chemogenic. The facies consists of beds of conglomerate or conglomeratic sandstone, generally 10–60 cm thick, separated by a few centimetres of mudstone or siltstone. The coarser beds vary laterally in thickness, and adjacent beds may converge to wedge out intervening finer-grained sediment. Flute casts, groove casts and load casts may occur on the base of coarser beds. Large-scale, tabular cross bedding and washouts are common. *Skolithos* may occur in the finer beds. The strong current activity which is indicated suggests a shallow-water, littoral environment, and this is confirmed by the fact that the facies occurs at the base of the transgressive Arenig succession.

The *Skolithos* Sandstone Facies is finer-grained and better sorted. Beds of sandstone are 10–30 cm thick and are again separated by a few centimetres of finer sediment. Large-scale cross bedding of both tabular and festoon types is common. The upper surfaces of the beds are often ripple-marked. *Skolithos* and *Planolites* are common and the top surfaces of beds are covered with worm trails and burrows. Deposition in shallow water is implied by the inorganic structures, and *Skilithos* indicates a littoral or sub-littoral environment.

The Oolitic Facies consists of ferruginous ooliths in a black matrix of mud or greywacke, interbedded with black, graptolitic shales. No trace fossils have been found. Ooliths accumulate today in a metre or so of water and are accepted as evidence of very shallow water.

The *Cruziana* Sandstone Facies consists of medium-grained sandstones a few cm to 1 m thick, with subordinate silt or mudstones a few centimetres thick. Large-scale symmetrical ripples and large-scale tabular and trough cross bedding are common. Washouts, flutes, grooves and load casts are rare. *Skolithos* is rare, but trilobite resting impressions (*Rusophycus*) occur and trilobite furrow tracks (*Cruziana*) are more common. The facies is believed to indicate deposition above wave base but in the sub-littoral zone.

The *Fodinichnia* Shaly Sandstone Facies consists of shales separating thin beds (about 1·5 cm) of fine sandstone or siltstone with shale laminae. Small-scale cross stratification is common, and feeding burrows are abundant. *Phycodes* is common in the fine sandstones and coarse silts and *Teichichnus* in the fine silts. Fairly quiet conditions are indicated.

The Silty Mudstone Facies consists of dark grey or black, sometimes graptolitic, mudstones with subordinate silty laminae. Some burrowing has taken place. The sediments are thought to have accumulated in quiet water, but it is only where many tens of metres of Silty Mudstone Facies succeed *Fodinichnia* Shaly Sandstone Facies that progressive deepening is indicated and it is safe to assume deeper water.

The Spiculiferous Chert Facies consists of white to yellow cherts interbedded with dark grey mudstones and shales. The cherts may show small-scale cross bedding, convolute lamination, and large-scale ripples, and they contain sponge spicules. The large-scale ripples may suggest deposition above wave base.

Finally, Crimes recognised a Volcanic Facies. This is of limited occurrence and consists of tuffs which may be either subaerial or marine, and pillow lavas which indicate marine conditions.

Some of the facies are believed to be depth-controlled and are taken to indicate a sequence from very shallow-water Conglomeratic Facies, Oolitic Facies and *Skolithos* Sandstone Facies, through *Cruziana* Sandstone Facies and *Fodinichnia* Silty Sandstone Facies, to the generally deeper-water Silty Mudstone Facies. The facies distribution and current directions suggest a shoreline not far to the northwest of Aberdaron and trending southwest to northeast.

In central Llŷn, basal Arenig conglomerates are exposed north and northwest of Abersoch. They are overlain by bioturbated grey and brown shales and similar sediments extend in a belt 1–2 km wide northeast of a line from Abersoch to Botwnnog. The sediments are probably over 200 m thick, and *extensus* zone faunas have been obtained by Nicholas (1915), Matley (1938) and Crimes (1969a). They are succeeded by grey and black, finely laminated shales with thin siltstone horizons of the *hirundo* zone.

Near Trefor, in extreme southwestern Arfon, previously unmapped sandstones and shaly sandstones are well exposed in the cliffs of Trwyn-y-tâl. Cross lamination and feeding burrows are abundant and the sediments probably represent the *Fodinichnia* Shaly Sandstone Facies. The sandstones are overlain by an oolitic and pisolitic ironstone consisting of ferruginous ooliths and pisoliths in a black mud matrix. The ironstone represents the Oolitic Facies, and is in turn succeeded by a series of hard, silty mudstones which probably represents the Silty Mudstone Facies. All three facies are probably therefore of Arenig age, although no diagnostic fossils have yet been obtained.

Further east, Arenig rocks are exposed on the western limb of the Ynyscynhaiarn Anticline near Criccieth, and on the eastern limb at Porthmadog. The basal member is a granule conglomerate or conglomeratic sandstone which rests on an erosional surface cut into underlying Tremadoc Slates. The basal sandstone, which has yielded chemogenic nodules formerly known as *Bolopora undosa*, is overlain by thinly bedded, fine-grained sandstones and siltstones with mudstone partings, all much bioturbated. A similar but generally thicker succession strikes northeastward from Penrhyndeudraeth to Blaenau Ffestiniog, where the strike swings to the southeast and the succession can be traced towards Arenig. It is interesting to note that Lynas (1973) thought the basal conglomeratic sandstones might be deltaic and that the assumed bioturbated nature of the overlying flaggy sediments might, in fact, be due to soft-sediment deformation as a result of the sudden expulsion of pore space water.

Arenig rocks are also known from Cwm Pennant. The Upper Pennant Quartzite, thought by Shackleton (1959) to be either Upper Cambrian or Arenig, and by Roberts (1967) to be Llanvirn, has been shown by Crimes (1969b) to be of probable Arenig age due to the presence of the trace fossil *Phycodes circinatum* (Richter). Three other small inliers of Arenig strata occur towards the head of Cwm Pennant.

Arenig strata also crop out in Arfon and have been described by Elles (1904) and Greenly (1944). Elles described the Afon Seiont section at Caernarfon in a paper in which she introduced the Arenig zones of *D. extensus* and *D. hirundo*. This classic section is probably faulted at the base and is incomplete. It begins with about 130 m of interbedded, grey-black shales, siltstones and thin, often calcareous, fine-grained sandstones of the *extensus* zone, and is followed by about 200 m of rather similar, but generally finer-grained, rocks of the *hirundo* zone. Elsewhere in Arfon, Greenly (1944) has described a basal sandstone, occasionally with lenses of granule conglomerate, resting unconformably on members of the Arvonian and reaching 65 m in thickness. The sandstone is succeeded by about 250–330 m of black shales of the *extensus* zone. The *hirundo* zone again consists predominantly of black shales, but is of uncertain thickness, perhaps 100 m or so. It includes a thin, impersistent pisolitic and oolitic ironstone which indicates a temporary shoaling of the sea.

2. Llanvirn

In western Llŷn, rocks of the *bifidus* zone crop out at Mynydd Rhiw, where shales with bedded tuffs and an ignimibrite are seen. The volcanics thin rapidly to the north and soon die out along the strike. In central Llŷn the sediments of the *bifidus* zone are grey and brown shales with thin siltstones, whereas the overlying *murchisoni* zone is represented by grey and black shales.

In southwestern Arfon, grey and black *bifidus* and *murchisoni* shales are known to crop out near Clynnog, but they do not appear in southeastern Llŷn and western Eifionydd. They are last seen at Llanbedrog where, dipping northeast, they strike into the sea. Sporadic outcrops are seen along the southeastern margin of the Vale of Nantlle where grey, black and dark blue slates with thin, silty laminae dip southeast beneath the younger rocks of the Llwyd Mawr Syncline. They are continuous with the Maesgwm Slates of Snowdon. Thicknesses are probably of the order of 350 m, but the detailed structure is unknown. Similar lithologies and thicknesses are seen in Cwm Pennant. They extend to the sea near Criccieth on the western limb of the Ynyscynhaiarn Anticline, but they fail to appear on the northeastern limb as a result, it was claimed (Fearnsides 1910), of thrusting. Shackleton (1959) suggested that the beds are absent as a result of unconformity at the base of the Caradoc, but he went on to claim evidence of thrusting localised near the base of the overlying Caradoc. The beds are also missing further to the east and northeast, where Caradocian rocks certainly rest unconformably upon Arenig strata. Elsewhere the same blue, black and grey slates extend in a continuous belt from the head of Cwm Pennant away to the northeast, to Nant Ffrancon and beyond. Throughout the area of Llŷn and Snowdonia, the rocks remain remarkably constant in lithology, indicating uniformity of conditions of sedimentation. The slates and shales contain a graptolite fauna which, in a few favoured localities, is abundant; the silty laminae have yielded fragmentary trilobite remains. Volcanic horizons are known only from Mynydd Rhiw in southwestern Llŷn.

3. Llandeilo (?)

Greenly (1944) ascribed a 4 m thick pisolitic and oolitic ironstone, together with associated black slates cropping out at Llandegai, to the Llandeilo Series on the basis of the occurrence of the graptolites *C. scharenbergi* (Lapworth) and *G. teretiusculus* (Hisinger). In the same way, Matley (1938) recorded *G. teretiusculus* from central Llŷn and assumed the Llandeilo Series to be present; the claim was made on similar evidence from Snowdon by Williams (1927), and from Cwm Cywion by Williams (1930). In each case, however, the shelly fauna on which the series is defined is not present, and the situation is made more unsatisfactory by the controversial status of the *teretiusculus* zone. Thus Skevington (1969) has argued that the zone is of doubtful validity, whereas Toghill (1970) has taken an opposite view. What is certain, however, is that the record of *G. teretiusculus* does not prove the presence of the *teretiusculus* zone and it certainly does not allow the recognition of the Llandeilo Series.

The lithology of the *G. teretiusculus*-bearing strata is similar to the slates and shales of the underlying proven Llanvirn strata, and it seems that quiet, perhaps deep-water conditions persisted. Apart from graptolites, the only fossils in the blue, black and grey shales and slates of the Upper Arenig, Llanvirn and presumed Llandeilo rocks are obscure horny brachiopods, phyllocarids and uncommon trilobites. With the exception of the oolitic ironstones, which indicate temporary shoaling and the subaerial ignimbrites of Llanvirn age at Rhiw, the rocks probably accumulated below wave base.

4. Caradoc

(i) *The Base of the Caradoc*

The rocks of the Caradoc Series are very variable and contain extensive developments of volcanic rocks. They crop out over much of Snowdonia and central and eastern Llŷn. The base of the Caradoc is taken at or a little below the incoming of *N. gracilis*.

In Llŷn, the base is exposed at Llanbedrog where fossiliferous, coarse-grained feldspathic sandstones with a Soudleyan fauna rest on *murchisoni* zone blue slates. Further west, in the northern part of St Tudwal's, Arenig mudstones (the Llanengan Mudstones) which have yielded a fauna indicative of the zone of *D. extensus*, are overlain (according to Crimes 1969a) by the 20 m thick Hen-dy-Capel ironstones. These consist of beds of pisolitic ore up to 5 m thick separated by black shales, usually less than 1 m thick and which have yielded graptolites indicative of the zone of *N. gracilis*.

In central Llŷn, east of Garn Fadrun, an andesite rests on *G teretiusculus*-bearing shales and is itself overlain by fossiliferous sandstones which have yielded a Lower Longvillian fauna to Crimes (1969a). In eastern Llŷn, near Four Crosses, the base of the Caradoc succession is not seen, but the rocks are volcanics with interbedded sediments containing an Upper Longvillian fauna. A similar situation exists near Llanystumdwy, except that the fauna in the lowest sediments is Soudleyan. Further east on Llwyd Mawr, Llanvirn slates are overlain disconformably by the Llwyd Mawr ignimbrite sheet which is presumed to be of Caradocian age. It is not until the country east of Cwm Pennant is reached that fossiliferous Caradoc sediments are seen again. On the northwestern flank of the Ynyscynhaiarn Anticline they were thought by Shackleton (1959) to be thrust over Tremadoc sediments but, along the eastern side of Cwm Pennant the junction, now against Llanvirn slates, is thought to be disconformable. The junction strikes northeast from the head of Cwm Pennant across Snowdonia, being sometimes a disconformity, at others a strike fault. In southern and southeastern Snowdonia, east of the alluvial spread of the Afon Glaslyn, the Caradoc rests essentially disconformably upon various members of the Arenig and Tremadoc.

There is therefore evidence of an important break near the base of the Caradoc in Llŷn and Snowdonia because the oldest Caradoc rocks rest variously on strata ranging from Tremadoc to Llanvirn and possibly Llandeilo in age. The break seems to be of least importance in the north and northwest, and to increase in magnitude to the south and southeast, towards the Harlech Dome.

(ii) *The Central Snowdonia Succession*

The lowermost Caradoc sediments are generally slates and silty slates. They sometimes show a local development of oolitic ironstones, which is presumably indicative of shoaling, and sometimes they are cherty. Graptolite faunas, such as those at Moel Hebog, for example, indicate the presence of the zone of *N. gracilis*.

The earliest volcanic rocks are developed in the Moelwyn district. Here the Moelwyn Volcanic Group occurs close to the base of the Caradoc succession and extends southwest towards Penrhyndeudraeth and thence northwestwards from Tremadog, where it is known as the Glôg Volcanic Group. It comprises a group of strongly altered lavas, tuffs and agglomerates of originally andesitic to dacitic composition, but the rocks are often referred to in the older literature as keraptophyres and soda-rhyolites. A similar, but thicker group of volcanic rocks occurs at about the same horizon in the north near Nant Peris and extends discontinuously northeast to Nant Ffrancon. Ashy sediments towards the top of the Moelwyn Volcanic Group contain shelly faunas indicating a Costonian age.

At a higher horizon within the Caradoc the predominant slates give way to sandstones and a second group of volcanic rocks, the Capel Curig Volcanic Group, crops out around Capel Curig. They reach their greatest development on Y Glyder Fach. The rocks are mainly welded tuffs (that is, welded ignimbrites and welded ash flows) and coarse breccias of rhyolitic composition, but bedded and water-lain tuffs and fossiliferous sediments are present away from Y Glyder Fach. The faunal evidence indicates that the group is of Soudleyan age.

Generally, sandstones (which are often very rich in reworked volcanic material) dominate over slates at higher horizons in the succession, and then a third group of volcanic rocks is developed: the well known Snowdon Volcanic Group. This group from Snowdon was described by Williams (1927), where it begins with the Lower Rhyolitic Group overlying the Gwastadnant Grits. The basal member of the group is the rhyolitic ignimbrite sheet known as the Pitt's Head Flow. The sheet is generally strongly welded with pronounced eutaxitic structure and consists of at least two ignimbrite units. Locally, there is a nodular development near the base of each unit. It lies unconformably upon members of the Glanrafon Beds and, if the correlation of the Pitt's Head Flows with the Llwyd Mawr ignimbrite (as suggested by Shackleton 1959 and Roberts 1967) is correct, then it oversteps onto Llanvirn slates on Llwyd Mawr. Beavon (1963) has shown that 1700 m of Glanrafon Beds are missing beneath the volcanics southeast of Beddgelert. The Pitt's Head Flow is succeeded by about 500 m of rhyolitic tuffs on Snowdon itself, but in the east and southeast it is overlapped by the Llyn Dinas Breccias. This is a remarkable formation consisting of 170 m of unstratified breccias with fossiliferous, intercalated sandstones and slates of Soudleyan age. The blocks in the breccia range up to 1 m or so across, and the formation was thought by Beavon (1963) to be the result of volcanic collapse following magmatic doming, but elsewhere Rast (1961) regarded them as lahars (that is, the products of volcanic mudflows). The succeeding Lower Rhyolitic Tuff overlaps the Llyn Dinas Breccias and the Pitt's Head Flows onto Glanrafon sandstones and slates. The formation consists of welded and non-welded ignimbrites on Snowdon itself, but the ignimbrites pass upward and laterally to the southeast and southwest into bedded, water-lain tuffs and sediments. The fauna suggests a Lower Longvillian age. Members of this formation are historically interesting because on Y Lliwedd, they were thought by Dakyns and Greenly (1905) to have been deposited by *nuées ardentes*. In a sense, therefore, they were the first ancient ignimbrites to be recognised in Britain, or, for that matter, anywhere else in the world, although of course the term *ignimbrite* had not at that time been coined.

The various members of the Lower Rhyolitic Group were followed by the Bedded Pyroclastic Group. On Snowdon, the group reaches 400 m in thickness and consists mainly of green, bedded pumice tuffs of andesitic or basaltic composition. Blocks, bombs and lapilli are abundant. To the southwest on Moel Hebog, the formation, known here as the Middle Basic Group, consists of basic tuffs, agglomerates and basalt lavas. A shelly fauna in marine tuffs near the base indicates an Upper Longvillian age. East of Beddgelert, the Middle Basic Group is broadly similar to the Moel Hebog succession and at Dolwyddelan the formation, represented by basic tuffs and intercalated basic lava, has thinned to 270 m and further east passes into slates and calcareous sediments. Northwest of Y Glyder Fawr the formation is 130 m thick and consists of basic pumice tuffs with 65 m of intercalated basic lava.

The succeeding Upper Rhyolitic Group on Snowdon consists of 100 m of bedded lapilli tuffs of rhyolitic composition resting on an erosional surface cut in the underlying Bedded Pyroclastic Group. Similar rocks are seen on Moel Hebog where they reach just 10 m in thickness. To the east at Dolwyddelan, the tuffs are interbedded with slates and the formation is 210 m thick.

Younger rocks are seen only in the Dolwyddelan Syncline where over 30 m of graptolitic, black, pyritous slates are exposed. They represent the higher part of the zone of *D. clingani* (Williams and Bulman 1931) and mark the cessation of Caradocian volcanism.

(iii) *The Succession in the Conwy–Trefriw Area*

The base of the Caradoc succession in this area is not seen. The oldest rocks are poorly exposed, dark grey and purple slates with bands of coarse-grained feldspathic sandstone. The slates have yielded *Glyptograptus teretiusculus* near the Penmaen-mawr intrusion. The succeeding Conway Volcanic Group is about 1300 m thick. Elles (1909) and Stevenson (1971) regarded the lowest member as a series of flow-banded, flowage-folded, and sometimes autobrecciated rhyolite lavas, but the author accepts the view of Davies (1969) and interprets it as an intrusive sheet. The thickness varies from about 400 m at Conwy to about 900 m on Foel Lus. At Conwy the rhyolite is overlain by the Bodlondeb Ash Group which is about 20 m thick and contains a thin shale band from which Elles (1909) obtained *G. teretiusculus*. The group is thought to correspond to about 30–60 m of shales and tuffaceous shales in the country westsouthwest of Conwy. They are succeeded on Conwy Mountain by about 230 m of rhyolitic rocks which Elles regarded as rhyolitic lavas, whereas Stevenson (1971) regarded them as tuffs and equated them with the tuffs of the Gyrach Division which occur to the southwest of Conwy. Certainly much of the rock exposed on Conwy Mountain has the appearance of an agglomeratic rhyolitic tuff, but the tuffs are also reinforced by thin sills of flow-banded and flowage-folded rhyolite. The tuffs are succeeded by 150 m of sandstones of the Coetmor Ash Group at Conwy and by 80 m of sandstone of the Maen Amor Division to the southwest. The Conway Volcanic Group closes with up to 150 m of rhyolitic tuffs, some of which are welded. They probably represent the Capel Curig Volcanic Group.

The succeeding Glanrafon Beds are up to 460 m thick just southwest of Conwy, and consist of sandstones passing up into slates. Towards Trefriw they thicken to about 600 m as grey slates appear beneath the sandstones, whereas at Conwy itself the sequence is much condensed and has yielded an Upper Longvillian fauna to Wood and Harper (1962). The beds were included by Elles (1904) in the Coetmor Group, a member of the Conway Volcanic Group. Elsewhere, faunas indicate an age range from Harnagian to Longvillian, although the bulk of the strata are of Soudleyan age.

The overlying Crafnant Volcanic Group is broadly the equivalent of the Snowdon Volcanic Group. It is thickest in the south in the Trefriw area where it probably exceeds 500 m, but it thins northward to 350 m at Trecastell, and it is absent from Conwy. The group consists of acid and basic pyroclastics with interbedded sediments. Three divisions were recognised by Stevenson (1971). The Lower Division is up to 120 m in thickness and consists mainly of basic pumice tuff and rare pillow lavas in the north. These give way to welded rhyolitic tuffs in the south which continue to Trefriw. The Middle Division consists of rhyolitic tuffs and black slates, but locally basic pumice tuffs are also present. The Upper Division is marked by a persistent basic pumice tuff which locally is overlain by black slate or rhyolitic tuff.

The Crafnant Volcanic Group is overlain by the Llanrhychwyn Slates, a group of sooty black, pyritous slates which locally yield abundant graptolites, indicating the *clingani* zone. The slates are succeeded near Trefriw by the Trefriw Mudstone and Trefriw Tuff. The tuff is basic and sometimes agglomeratic and represents the youngest volcanic rocks in the area.

(iv) *The Llŷn Succession*

The Caradoc succession in Llŷn is even more variable than in Snowdonia and its study is further complicated by the restricted and isolated nature of the outcrops. In southwestern Llŷn, in the northern part of St Tudwal's peninsula, we have seen that the Caradoc rests unconformably upon Arenig strata. The Caradoc here consists of the Hen-dy-Capel Ironstones—pisolitic ironstones and intercalated shales, with a graptolite fauna indicating the zone of *N. gracilis*—succeeded by the Pen-y-Gaer Mudstones. The succession is thought by Crimes (1969a) to be terminated by a thrust.

In the east and northeast of Central Llŷn, the succession begins with an andesite where, east of Garn Fadrun, it is seen resting on *G. teretiusculus*-bearing shales and is succeeded by fossiliferous sandstones from which Crimes (1969a) obtained a Lower Longvillian fauna and hence the Costonian, Harnagian and Soudleyan stages are absent.

At Llanbedrog fossiliferous, coarse-grained, feldspathic sandstones rest disconformably on soft, blue slates of the zone of *D. murchisoni*. They pass up into richly fossiliferous mudstones and both sandstones and mudstones have yielded a Soudleyan fauna of trilobites and brachiopods. They are succeeded (according to Fitch 1967) by 165 m of hornblende-andesite lavas, 80 m of bedded tuffs and sediments, 100 m of hornblende-dacite lavas, and then by 200 m of bedded tuffs, breccias, agglomerates and lavas of rhyo-dacitic composition. About 600 m of welded and non-welded ignimbrites rest unconformably upon the rhyo-dacites and the ignimbrites are in turn overlain by 225 m of variable rocks called the Penarwel Drive Beds. They include acid lavas, tuffs, sandstones, mudstones and black shales. The sandstones contain an Upper Longvillian fauna and the black shales yield graptolites suggesting the zone of *M. multidens*. These variable beds are overlain by 80 m of black shales (the Nod Glas Beds) which contain graptolites, indicating the zone of *D. clingani* and which terminate the Caradoc succession in this locality.

In the Chwilog area, a badly exposed volcanic succession is claimed by Tremlett (1965) to consist of more than 550 m of mainly bedded, rhyolitic tuffs with lesser agglomerates, ignimbrites and andesitic lavas. Mudstones are intercalated with the volcanic rocks towards the top of the succession and the volcanic formation is succeeded by fossilferous grey shales and siltstones with an Upper Longvillian fauna. Most of the rocks described as bedded, rhyolitic tuffs are regarded by the author as rhyolite sheets because the majority show flow-banding and flowage-folding, and pyroclastic features are absent. As a consequence, the volcanic sequence is now considered to be much thinner than has been claimed.

Further to the east, just to the north of Llanystumdwy in the beds of the Afon Dwyfor and Afon Dwyfach, blue-grey shales, with a possible Harnagian shelly fauna, underlie a thick sequence of bedded and welded tuffs of largely rhyolitic composition. They are succeeded by black shales, flaggy beds and grey mudstones with a shelly fauna spanning the Longvillian. There is then a pronounced facies change and the succeeding beds are a succession of black, flaggy, graptolitic mudstones and shales, the first 15 m of which represent the zones of *D. clingani* and *P. linearis*. Harper (1956) suggested the graptolite faunas present at Llanystumdwy might also indicate the presence of the zone of *D. complanatus*.

In northeastern Llŷn, near Nefyn and Llanaelhaearn, slates, mudstones and siltstones of presumed Llanvirn age are overlain by a volcanic formation which begins with andesite lavas, followed by a group of rhyolitic tuffs and agglomerates with intercalated, fossiliferous sediments. The fauna indicates an Upper Longvillian age for the rocks. The volcanic formation is overlain in turn by Upper Longvillian calcareous shales and finally by unfossiliferous, dark grey slates. At Clynnog Fawr a rather similar succession is developed. The volcanic rocks rest on Llanvirn slates and again andesite lavas are succeeded by rhyolitic rocks, claimed by Tremlett (1964) to be air-fall tuffs and ignimbrites. The succession is terminated by about 130 m of black slates. Here once again, however, in view of the presence of flow-bands and flowage-folds and the absence of bedding or eutaxitic structures in the rhyolitic rocks, the author prefers to regard the rhyolitic rocks as either lavas or intrusive sheets.

5. Ashgill

Ashgillian rocks are known from the Llanbedrog area, from the area north of Llanystumdwy, and from the area between Conwy and Trefriw.

At Llanbedrog the Nod Glas Beds are succeeded by the Crugan Mudstones, which are over 300 m thick. They consist of grey, blocky mudstones, sometimes silty and sometimes with calcareous nodules. The lower 200 m contain small shelly fossils representing the *Phillipsinella parabola* zone. At Llanystumdwy along the Afon Dwyfach, rather similar grey mudstones have yielded a rich, fragmentary trilobite fauna which again represents the *P. parabola* zone. The beds are particularly interesting because Harper (1956) records a graptolite-bearing horizon within the mudstones which probably indicates the presence of the *Dicellograptus complanatus* zone.

Ashgillian rocks occur extensively between Conwy and Trefriw. At Conwy, Elles (1909) separated grey mudstones into the Bodeidda and Deganwy Mudstones. Further south the mudstones are about 350 m thick and consist of a monotonous sequence of bioturbated strata with thin, nodular limestones. Near Trefriw, the rocks are cleaved and are known as the Grinllwm Slates. At Conwy the mudstones are succeeded by the Conway Castle Grit consisting of about 50 m of calcareous turbidite greywackes and shales which yield a Hirnantian fauna.

6. Conditions of Accumulation of the Ordovician Rocks

The earliest Arenig sediments are usually coarse-grained and show a progressive upward fining to shales and mudstones. The fauna and sedimentary structures suggest a change from littoral, through sub-littoral, to deeper-water, quiet conditions with most of the sedimentation taking place below wave base. A shoreline is indicated away to the northwest. Contemporary marine volcanism is indicated by the presence of basic pillow lavas, but evidence of temporary shoaling is provided by the oolitic ironstones, and it may be that some of the bedded tuffs were formed under short-lived subaerial conditions. Generally, however, conditions of deeper, quiet water appear to have continued throughout much of the Arenig and the Llanvirn. An exception to this occurred at Rhiw where, during the Llanvirn, a rapid and local build-up of marine volcanics resulted in temporary emergence and the formation of rhyolitic ignimbrites. The *G. teretiusculus*-bearing slates indicate a continuation of the same generally quiet marine conditions with deposition taking place below wave base.

A glance at the 1:25 000 geological map of Central Snowdonia indicates that uplift and erosion, followed by re-submergence, then occurred because the basal Caradoc rocks rest on various formations ranging in age from Tremadoc to Llanvirn. It is also clear that the same events took place in Llŷn. The Caradoc was marked by intense volcanic activity and consequently by a complicated palaeogeography as volcanic centres waxed and waned. A number of important breaks in the succession occur and unconformities are developed at several horizons within the Caradoc, indicating periods of emergence and erosion. Such breaks are seen, for example, at the base of the Pitt's Head Flows, at the base of the Llyn Dinas Breccias, and at the base of the Lower Rhyolitic Tuffs. Extensive subaerial accumulation of volcanic rocks is indicated by the various ignimbrite sheets but the bulk of the volcanics and probably all the sediments accumulated under marine conditions. An important feature of the Caradoc succession is the large volume of sandstones of largely volcanic origin. The bulk of the constituents probably result from volcanic activity within, rather than outside, the basin of accumulation. A second important feature is the very considerable increase in thickness, towards the southeast, of the Caradocian sediments (Glanrafon Beds) underlying the Snowdon Volcanic Group. The sediments are absent from Llwyd Mawr, about 1250 m thick on the western flank of Moel Hebog, 1750 m on the eastern flank, and about 2750 m thick at Croesor. The centre of the subsiding basin was therefore southeast of Croesor.

Presumably Llŷn and Snowdonia formed part of an active volcanic island arc throughout the Caradoc, but volcanic activity appears to have largely ceased towards the close of the Caradoc and, apart from the outburst represented by the Trefriw Tuff, the late Caradoc and Ashgill sediments suggest a return to more uniform marine conditions.

Fitton and Hughes (1970) proposed that the Caradocian volcanic rocks of North Wales accumulated in an island arc environment above a southeasterly dipping Benioff zone, and that the associated oceanic trench was situated in Scotland, running northeast to southwest through Girvan. Certainly the Lower Ordovician succession suggests accumulation on a labile marine shelf. The development of volcanic conditions began on a very small scale in the Arenig with pillow lavas and very small quantities of bedded tuffs in Llŷn; the volcanics of Rhiw indicate an increase in volcanic activity with the emergence of subaerial volcanoes and the first establishment of the volcanic arc. However, the strong development of volcanic arc conditions began with the basal Caradoc unconformity and the eruption of the Moelwyn Volcanic Group, and the peak was reached in the Soudleyan and Longvillian with the eruption of the Capel Curig, Crafnant, Snowdon, and Llŷn Volcanic Groups. It is at this time that the great variety of sediments and volcanics, and the rapid facies changes, indicate the maximum complexity of palaeogeographical conditions. Presumably the Benioff zone ceased to be active at the close of the Caradoc and non-volcanic shelf sea conditions were re-established during the end of the Caradoc and the Ashgillian.

7. Intrusive Igneous Rocks

A large number of minor and medium-sized igneous intrusions cut the Ordovician and the Cambrian strata. They include dykes, sills and boss-like masses, and the rock types range from ultramafic, through mafic and intermediate, to felsic. Many can be shown to be of probable Ordovician age.

(i) *Dolerites*

Dolerite sills are particularly abundant in the Caradoc rocks of Snowdonia but they are also emplaced, though perhaps less abundantly, throughout the rest of the Ordovician and the Cambrian. When fresh they consist of augite, labradorite and iron ore but they are usually altered, mainly by low-grade regional metamorphism associated with the folding, so that now they usually consist of a low greenschist facies assemblage which includes albite, epidote, chlorite, actinolite, sphene, calcite and leucoxene, together with relict plagioclase and augite. Pumpellyite, prehnite and stilpnomelane have also been recorded. Textures are typically ophitic and igneous textures remain recognisable except in the most strongly cleaved rocks. Grain sizes range from medium to very coarse, the coarser rocks being gabbros in the strict sense rather than dolerites. In some coarse-grained varieties augite and plagioclase crystals may reach over 2 cm in length.

The age of emplacement of the sills has been a topic of discussion for a long time. Thus Williams (1927), for example, described them as pre-cleavage; Williams and Bulman (1931) as both pre- and post-cleavage; and Shackleton (1953) as late Caledonian. Bromley (1969), in a review of acid plutonic activity of North Wales, pointed out that dolerite sills are best developed in and immediately below the Ordovician volcanic rocks, and that they are absent from higher strata. This latter fact at least is readily understandable, if only because erosion has left us with very few and very restricted areas of Ashgillian and Llandovery strata. The frequent occurrence of dolerite sills in older strata is less readily explainable. However, in Llŷn a number of sill-like dolerite intrusions are seen to be emplaced mainly within the Lower Ordovician, but again this is probably because a large part of Llŷn is occupied by only Lower Ordovician rocks.

Dolerite dykes (which are petrographically identical to the dolerite sills) are also abundant, although they seldom run for more than 2 km and are usually much less in length. Thus small dykes occur in Upper Cambrian and Lower Ordovician strata in the southeast between Deudraeth and Ffestiniog, where they trend between east–west and east-northeast–westsouthwest. They are also seen cutting the Cambrian Slate Belt where the trend is between east–west and westnorthwest–east-southeast. Another good development is around Tryfan and Llyn Ogwen, where the trend is again westnorthwest–eastsoutheast. It is evident, despite claims to the contrary by Evans (1968), that in this last locality the dykes fill tensional features associated with the main period of folding, and it may be that this explanation also holds for the other occurrences. It is also worth bearing in mind that the dolerite dykes of Llŷn, although not numerous, also show a general east–west trend.

The effects of very low grade and low grade regional metamorphism associated with the end-Silurian to Devonian folding are well seen in the basic igneous rocks. Thus dolerites and basic tuffs in the area immediately southwest of Conwy show alteration of plagioclase to albite, prehnite and pumpellyite and of clinopyroxene to pumpellyite (Jenkins and Ball 1964). Similar very low grade metamorphic assemblages are developed in Llŷn in basic rocks west of Abererch and the rocks in both areas belong to the prehnite–pumpellyite facies and the pumpellyite–actinolite facies. Elsewhere dolerites and basic tuffs carry typically low greenschist facies minerals including clinozoisite, actinolite, stilpnomelane and biotite.

(ii) *Composite Sills*

Composite sills from 2–100 m thick were described by Beavon (1963) from southeast Snowdonia. Each intrusion usually consists of a thick quartz–feldspar porphyry or felsite enclosed in a thin envelope of quartz–dolerite or dolerite. They are usually cleaved, but are cut by later dolerite dykes. They were thought by Beavon to be coeval with the Snowdon Volcanic Group. On the other hand, Bromley (1969) believed they were the result of mafic magma having remobilised older quartz–latite intrusions which had been emplaced previously at lower horizons and which were related to the Moelwyn Volcanic Group. Certainly, the composite sills have a limited areal extent, being confined to the south-eastern limb of the Arddu Syncline.

(iii) *Layered Mafic Intrusions*

Cumulate rocks, including hornblende-picrites, hornblende-gabbros, leuco-gabbros and ore-rich gabbros, have been described from Rhiw in Llŷn by Hawkins (1970) and their geochemistry has been described by Cattermole (1969). Three thick, sill-like masses of rhythmically banded rocks, which are parts of a formerly continuous mass, are emplaced in Llanvirn sediments. The southernmost, Mynydd Penarfynydd, is 105 m thick and shows the best sequence of layered cumulate rocks with an extensive development of picrites. Above the chilled marginal facies is a zone of metadolerite and gabbro in which the plagioclase is now albite as a result of low-grade Caledonian metamorphism. Above this zone is a series of hornblende-picrites containing cumulus olivine and chrome-iron spinels enclosed, commonly, in intercumulus brown hornblende or, rarely, in clinopyroxene or plagioclase. Some strikingly honeycomb-weathered picrites are due to differential weathering of intercumulus plagioclase on the one hand, and intercumulus ferro-magnesian minerals on the other. Rhythmic layering is developed with individual bands extending laterally for at least 15 m. The picrites are sharply overlain by a rhythmically banded leuco-gabbro which consists largely of cumulus clinopyroxene and ophitic plagioclase, with a small amount of intercumulus hornblende. The upper layers of the leuco-gabbro are often irregular and sometimes folded. The unit is overlain by a pegmatitic gabbro of very coarse grain size. Then come highly altered hornblende-gabbros, still with cumulate textures and abundant prehnite after plagioclase. This is followed by an ore-rich gabbro, still with clear cumulate textures. It is overlain by a metadiorite, devoid of cumulate textures and thought to be close to the original roof. The highest exposed rock type is a metagranophyre consisting largely of a granophyric intergrowth of quartz and albite. Mynydd-y-Graig consists mainly of layered hornblende-gabbros and Mynydd Rhiw itself solely of hornblende-gabbro. The rocks were derived from a relatively water- and alkali-rich mafic magma which was probably emplaced during the Llanvirn.

(iv) *Felsitic Minor Intrusions*

Felsitic minor intrusions are abundant, particularly within the Caradocian rocks. They include porphyritic and non-porphyritic types; they range from quartz–latite to rhyolite in composition; and they include dykes, sills and domes. Many of the rocks show devitrification textures in thin section and represent former obsidians. They often show the development of flow-banding, flowage-folding and autobrecciation, and some show remarkable development of nodular and spherulitic textures. Domes range up to 1 km across, but are commonly less. Dykes are usually thin, being of the order of less than 10 m across, but sills may be up to 100 m thick, although again commonly less. It is generally accepted that the felsitic intrusions are related to the Caradocian volcanic rocks and are essentially sub-volcanic intrusions. There are very strong chemical and mineralogical similarities between these intrusions and the more extensive volcanic sheets and some can be shown, on field evidence, to be contemporary with the volcanic rocks.

(v) *Microtonalites, Microgranites, etc*

These rock types occur as boss- and stock-like intrusions as well as thick sill-like and laccolithic masses. The bosses reach up to 4 km in diameter and the laccolithic sheets similarly may be up to 4 km across and exceed 500 m in thickness. The rocks range in composition from calc-alkali to alkali, and include tonalite, granodiorite, granite and soda-granite. The rocks are commonly porphyritic with a granophyric, medium-grained matrix. Some, such as the Tan-y-grisiau granite, have been shown to have been emplaced prior to the main period of folding and cleavage formation, but others, such as the intrusions of northern Llŷn, are regarded by Tremlett (1962, 1964) as having been emplaced after the main period of Caledonian folding. The better known intrusions include the Tan-y-grisiau microgranite, the riebeckite-microgranite of Mynydd Mawr, the Bwlch y Cywion microgranite, the Llanbedrog granophyre, the Garn Fadrun granophyre and the riebeckite-microgranite of Mynytho. The larger intrusions have developed aureoles of contact metamorphism with hornblende-hornfels facies rocks originally developed against the intrusions.

SILURIAN

Silurian rocks are present only in southeastern Llŷn and the Conwy district. They crop out in the western bank of the Dwyfor immediately north of Llanystumdwy, where the rocks overlie green and grey Ashgillian slates and mudstones and dip at low angles to the west. The succession begins with 20 m of black to greyish black pyritous mudstones. They are colour-banded due to the presence of very thin silt laminae and are only feebly cleaved. Monograptids are abundant and preserved in pyrite in high relief. The mudstones are succeeded by 30 m of blue-black and blue-grey shales with thin siltstone partings. Monograptids are again abundant. The rocks are of Llandovery age and include representatives of the *crispus* zone.

In the Conwy area the grey mudstones of the Ashgillian are succeeded by the Gyffin Shales, some 90 m thick. They are grey shales and silts with thin black mudstone laminae. The graptolite fauna indicates they are Llandovery in age. They are succeeded by turbidite siltstones, sandstones and intercalated shales, the Benarth Flags and Grits, the graptolite fauna of which indicates a Lower Wenlock age. The sandstone members thicken and coarsen upward. They are typical turbidite sandstones, often graded and with a wealth of sole markings.

Quiet marine conditions of accumulation, below wave base, are indicated for the Llandovery of both Llanystumdwy and Conwy. Turbidity currents began to invade the Llandovery sea towards the close and their activity gradually increased during the Wenlock in the Conwy area.

CARBONIFEROUS

Carboniferous rocks fringe the mainland shore of the Menai Straits between Bangor and Caernarfon. They are bounded on the landward side by the Dinorwic Fault and occupy a syncline. The succession is seen resting unconformably upon Ordovician (Arenig) shales, and is summarised in table 2.

The Basement Formation begins with a member called the 'Loam-Breccia' by Greenly (1928). It was interpreted as a talus overlying a loess. The blocks constituting the talus are locally derived from the Mona Complex. The rock is red-brown in colour and contains abundant ferric oxide pisoliths. It is a pisolitic ironstone with sporadically distributed, locally derived, angular blocks of the Mona Complex. The member is succeeded by conglomerates with sandstones, shales and thin limestones. Fossils present are mainly worm tubes and casts, together with plant remains. The proven D_1 beds are mainly fine-grained limestones (micrites) at the base, overlain by bioclastic limestones (biosparites) which contain thin lenses of sandstone and conglomerate. Productids characterise the micrites, whereas the biosparites are rich in corals. The debris of the biosparites consists, in order of importance, of fragments of crinoids, foraminifera, brachiopods and ostracods. The D_1 zone is over-

Table 2. The Carboniferous succession.

	Formation	Thickness (m)	Zone
Upper Carboniferous	Red Measures	720	
Lower Carboniferous	Cherty Formation	50	D_3
	Limestones with sandstones	65	
	Limestones with sandstones and shales		
	Pleurodon Mudstone Formation	200	D_2
	Limestones with sandstones and shales		
	Limestone with conglomerate and shale	30	D_1
	Basement Formation: Conglomerate	30	?
	'Loam-Breccia'	12	

lapped westward by rocks of the D_2 zone, which again tend to be micrites at the base succeeded by biosparites. Calcareous muds and shales are developed in the middle of the zone. The D_3 strata are mainly biosparites and a 30 m oolitic limestone is developed at the base. A few thin sandstones and conglomerates are included in the zone. The limestone sequence ends with the Cherty Formation which comprises stratified, thinly bedded cherts with interbedded cherty limestones. In the cherty limestones the chert is usually concentrated around Productid shells.

The Red Measures occupy a thin sliver of very poorly exposed ground between Porth Dinorwic and Caernarfon. Comparable rocks on Anglesey overlie Productive Measures and so the Red Measures are probably referable to the Upper Coal Measures. Near Caernarfon, about 20 m of plane-bedded red marls with occasional grey-green beds, and a few pebbly beds are seen. The succession includes a spectacular conglomerate consisting of pebbles of andesite, quartzite and felsite up to 20 cm across, held in a red, sandy matrix.

Although the Carboniferous rocks have a very restricted outcrop, it is clear that during much of Lower Carboniferous time the area was land and of considerable local relief. The earliest D zone sediments may be subaerial, but succeeding marine sediments are in no way unusual and consist of the customary carbonate shelf deposits. The Upper Carboniferous Red Measures are subaerial; they rest unconformably upon older rocks and bring the stratigraphical succession of Snowdonia and Llŷn to a close.

TERTIARY

Tertiary rocks are represented by a number of basic dykes which trend towards the northwest quadrant and range in thickness from a few centimetres to 40 m. Mineralogically, they consist of labradorite, clinopyroxene, olivine, iron ore and sometimes analcime, and have affinities with alkali-olivine basalt. Texturally, they range from basaltic to gabbroic, but most are doleritic. They weather rapidly and often occupy eroded slots in the hill country. Fitch *et al* (1969) regard the age of the dykes as 61 ± 3 million years.

2. Structure

INTRODUCTION

There has been no attempt at a synthesis of the structural evolution of Snowdonia and Llŷn since the important discussion of North Wales as a whole by Shackleton (1953). A number of subsequent publications have dealt with certain aspects of the structure, amongst which an historical essay by Bassett (1969), and a discussion of the relationship between Ordovician structure and volcanism by Rast (1969), are worthy of mention. The structure of North Wales was also considered in a wide-ranging review of the para-tectonic British Caledonides by Dewey (1969b).

It has generally been agreed, or at least implied in the past that the major structures in the Mona Complex are Precambrian in age, and that the major structures in the Lower Palaeozoic rocks are end-Silurian to Devonian in age (see, for example, George 1961). Radiometric dating tends to support this view. Thus, for the Coedana granite of Anglesey and its metamorphic aureole, we find ages of 580–600 million years (Moorbath and Shackleton 1966), and ages of up to 596 ± 15 million years for various other Mona Complex rocks. These provide minimum ages for the granite emplacement and the associated high-grade metamorphism of the Mona Complex, that is, the latest Precambrian, but presumably the low-grade greenschist and lawsonite–glaucophane metamorphism, together with the main phase of deformation, is older. Dates obtained from the Lower Palaeozoic rocks of Snowdonia and Llŷn suggested to Fitch *et al* (1969) that the main deformation of these rocks occurred 390–430 million years ago; that is, end-Silurian to Devonian. It would therefore seem that deformation and low-temperature, high-pressure metamorphism of the Mona Complex may have culminated more than 600 million years ago and prior to the accumulation of the Arvonian volcanic rocks; that the period of granite emplacement and high-temperature, low-pressure metamorphism quickly followed; and that a second major deformation of the Precambrian plus the Lower Palaeozoic rocks occurred at about the end-Silurian to Devonian.

Examination of the geological map of Snowdonia and Llŷn indicates, however, that a major unconformity exists at the base of the Ordovician. The generally accepted view is, nevertheless, that the break is not of orogenic proportions, but rather that it is the result of end-Tremadoc uplift and erosion, followed by subsidence (see George 1961). This interpretation is questionable because the Arenig strata come to rest with flagrant unconformity on strongly deformed amphibolites and migmatitic gneisses of the Mona Complex in western Llŷn, and clearly several thousands of metres of Precambrian and Cambrian strata were removed by erosion prior to the deposition of the Ordovician sediments. Furthermore, there is no really strong stratigraphical evidence—certainly none that is apparent in the geological maps of North Wales—in support of a pre-Arvonian major deformation of the bedded members of the Mona Complex. The evidence, such as it is, consists of a small patch of conglomerate of *presumed* Cambrian age which rests on beds of the Gwna Group at Trefdraeth on Anglesey, and a small patch of ignimbrite of *presumed* Arvonian age, which again rests on beds of the Gwna Group, at Holland Arms on Anglesey.

The situation is therefore that the geological evidence indicates two periods when major movements (probably of orogenic proportions) took place, namely, end-Silurian to Devonian and end-Tremadoc. However, when evidence from radiometric dating studies is also considered, a third and earlier period of deformation and associated metamorphism is indicated immediately prior to the accumulation of the Arvonian.

Within the last 20 years, however, various authors have made interesting and compelling claims for a measure of essential continuity across one or more of these periods. Thus Shackleton (1956) claimed that in western Llŷn there is continuity between the gneisses on the one hand, and the Bedded Series of the Mona Complex on the other, and that, by implication, both groups of rocks have suffered the same metamorphic and structural history. Shackleton (1969) has further suggested that the topmost members of the Bedded Series of the Mona Complex, the Gwyddel Felsitic Beds of Llŷn and their equivalents on Anglesey, the Fydlyn Felsitic Beds, may perhaps pass without any important break into the acid volcanic rocks of the Arvonian. He pointed out that there is a progressive change towards the top of the Mona Complex into a sequence of bedded acid volcanics with a few included ignimbrites, and that the younger Arvonian consists predominantly of ignimbritic volcanics. Furthermore, Wood (1969) has demonstrated that at Llyn Padarn, the Arvonian passes upwards conformably into undoubted Cambrian rocks. Important stratigraphical breaks are well documented within the Cambrian succession, but they express themselves merely as disconformities and in the present context can be regarded as of little structural importance. We thus have claims for an essential continuity from the gneisses and Bedded Series of the Mona Complex, though the Arvonian, to the end of the Cambrian; claims which appear to conflict with the radiometric evidence as well as some of the field evidence.

The apparent conflict can be resolved if we accept a view rather similar to that put forward by Shackleton (1969) in his answer to the discussion following his paper on the Precambrian of North Wales. Shackleton appeared to emphasise that regional metamorphism occurred successively as temperatures waxed and waned, and that whilst at depth

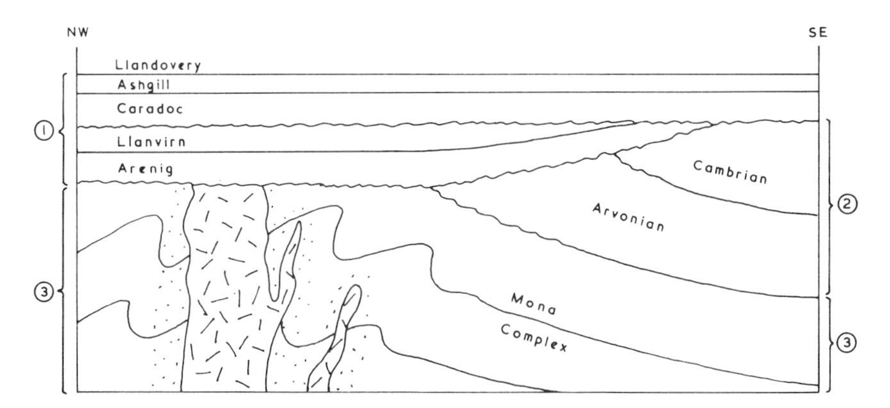

Figure 6. Diagrammatic representation of the relationships between the major rock groups.

regional metamorphism was taking place, this would necessarily be synchronous with other events taking place at higher structural levels. It is possible to reconcile the apparently conflicting evidence in a diagrammatic summary of the relationships between the major rock groups, as shown in figure 6.

This results in three groups of rocks which generally should show different structural histories and therefore different structural characteristics. We should be able to distinguish the following.

1. The Arenig–Llandovery rocks deformed primarily at the end-Silurian to Devonian.

2. The Arvonian–Cambrian rocks deformed firstly by the end-Tremadoc movements, and secondly during the end-Silurian movements.

3. The Mona Complex rocks, primarily deformed and metamorphosed before the accumulation of the Arvonian (on radiometric evidence), but deformed to some extent again by the end-Tremadoc movements, and finally by the end-Silurian to Devonian movements.

PRECAMBRIAN STRUCTURES OF THE MONA COMPLEX

Little is known in detail about the structure of the Mona Complex in Llŷn, largely due to the small area the rocks occupy, coupled with the very restricted areas of actual outcrop. The earliest structure seen is a pressure solution cleavage, S_0, which is sporadic in its occurrence and is parallel, or sub-parallel, to bedding. No related folds have yet been recognised. It has been shown by Shackleton (1956) that the dominant structures have a Caledonoid trend, face upwards and are slightly overturned to the southeast. They are the earliest structures recognisable and therefore constitute the F_1 Precambrian folds. In detail (see figure 12), a fold-pair can be traced west of Aberdaron for about 7 km. The pair has a wavelength of about 500 m, and plunges of associated minor structures are generally to the southwest at angles ranging from 0–70°. The more westerly member of the pair, a syncline, can be traced from near Dinas Bach to the coast at Trwyn Maen Melyn. Another major syncline can be recognised northeast of Aberdaron, the core of which is occupied by gneisses, but elsewhere F_1 folds are less readily recognised. In rocks of suitable composition and low metamorphic grade, the F_1 structures are accompanied by an axial-planar slaty cleavage, S_1, of Caledonoid trend and steep northwesterly dip.

Table 3. Correlation of some Precambrian structural elements of Llŷn and Anglesey.

Llŷn		Anglesey	
F_1	Caledonoid trend; axial surfaces dip steeply northwest	F_1	As Llŷn
S_1	Slaty cleavage in suitable lithologies	S_1	As Llŷn
	Not recognised	F_2	Caledonoid trend; axial surfaces gently inclined
		S_2	Axial-planar crenulation cleavage may be intense and obliterate S_1
F_2	Trend WNW–ESE; axial surfaces steeply inclined	F_3	Trend NW–SE; axial surfaces steeply inclined
S_2	Axial-planar crenulation cleavage in suitable lithologies	S_3	Axial-planar crenulation cleavage in suitable lithologies

A set of later structures, the Precambrian F_2 structures, deforms both bedding and S_1. The F_2 folds trend northwest with more or less upright axial surfaces, and are often accompanied by an axial-planar crenulation cleavage in rocks of suitable lithology. Their influence is well displayed at Mynydd Anelog, where F_2 structures deform the F_1 fold-pair and cause the F_1 axial traces to run north–south for about 1 km. These later structures also cause variations in the plunges of F_1 linear structures and are also largely responsible for the unusual outcrop pattern of the Gwyddel Felsitic Beds.

It should be pointed out that over many parts of the outcrop of Mona Complex rocks on Anglesey, a set of folds is interposed between the first structures, which are similar to the F_1 structures of Llŷn, and a third set of structures, which are similar to the F_2 structures of Llŷn. These additional structures have a broadly Caledonoid trend, but the axial surfaces range from moderately inclined to near horizontal in attitude, and the folds are commonly sideways-facing. It is probably safe to correlate the structures over such a short distance as in table 3.

THE END-TREMADOC STRUCTURES

Various authors have considered the magnitude of the sub-Arenig unconformity, and George (1961) illustrated it with a map which showed that the unconformity becomes stronger to the north and west of the Harlech Dome. He pointed out that uplift and erosion led to the removal of about 5000 m of mainly Cambrian rocks in Llŷn, whereas virtually nothing was removed in the vicinity of the Harlech Dome. The mechanism of the uplift is usually assumed to be vertical movements along major faults with a Caledonoid trend, such as the Nefyn Fault in Llŷn and the Dinorwic Fault bounding the Bangor Ridge. However, the situation is more complicated than the reconstructions of George and Shackleton suggest. Thus, for example, Arenig rocks have been identified at Abersoch resting with no discordance on high Tremadoc sediments; furthermore, within a distance of 4 km to the southwest the Arenig has overstepped to the Lower Cambrian, and 8 km west of Abersoch onto high-grade metamorphites of the Mona Complex. It seems unlikely that this was accomplished without the generation of large-scale folds in the sub-Arenig strata and, indeed, the map of Nicholas (1915) provides one of the best indications of the probable presence of the limb of one such fold. It remains true, however, that it is not possible to outline completely such a fold with certainty on the geological map nor to demonstrate convincingly the presence of such folds on the ground.

It is probably worth emphasising once again the fact that the geological evidence indicates only two periods of extensive deformation, namely, end-Silurian to Devonian and end-Tremadoc, and that the immediately pre-Arvonian deformation of the Mona Complex is indicated principally by radiometric evidence. If the radiometric evidence has been misinterpreted here, then it may well be that the deformation of the Bedded Series of the Mona Complex, the Arvonian and the Cambrian are all principally end-Tremadoc!

THE END-SILURIAN TO DEVONIAN STRUCTURES

It is generally accepted that the main fold architecture was produced by movements of orogenic proportions which occurred around the close of the Silurian period (see, for example, Shackleton 1954, George 1961), but the evidence for the timing of this event lies largely outside the area of Snowdonia and Llŷn. Briefly, in the Llangollen Syncline, Ludlovian strata are affected by the movements which affect the Lower Palaeozoic rocks of Snowdonia and Llŷn; and on Anglesey Old Red Sandstone facies rocks of presumed Lower Devonian age rest unconformably upon folded Ordovician and Precambrian rocks. Since Ordovician and Silurian rocks were folded together, the movements must therefore be about end-Silurian to Lower Devonian in age.

More recent work by Helm *et al* (1963), Roberts (1967) and Crimes (1969a) has shown that the main period of deformation can be subdivided into a number of phases. Thus Roberts (1967) distinguished and described three sets of structures, each of which was ascribed to a distinct phase of the deformation. The first phase, F_1, gave rise to the main fold architecture and produced an axial-planar slaty cleavage, S_1, in rocks of suitable lithology, such as pelites and many tuffs; folds trend approximately northeast–southwest and have axial surfaces which are vertical or steeply inclined to the northwest. The effects of the second phase, F_2, are uncommon and sporadically distributed. They consist of rare, rather open, sideways-facing folds with gently inclined or near-horizontal axial surfaces. The folds are developed in both bedding and S_1, and trend necessarily northeast–southwest. Rarely, a weak crenulation cleavage or sometimes a fracture cleavage, S_2, may be present in rocks of suitable lithology. S_2 bears an axial-planar relationship to the F_2 folds. The effects of the third phase, F_3, are more common than F_2 and sometimes are strongly developed as, for example, near Trefor in Llŷn. They consist of folds developed in bedding, S_1 and, very rarely, S_2. Folds trend from northwest–southeast to north–south, and commonly plunge steeply to the northwest quadrant. Axial surfaces are upright or steeply dipping and an axial-planar crenulation or fracture cleavage is commonly present.

A further contribution to the structure of Central Snowdonia has been made by Shackleton (in Beavon 1963), Bromley (1969) and Rast (1969). Shackleton suggested that some of the folds were of mid-Caradocian age and of volcano-tectonic origin. Briefly, it is claimed that the Arddu Syncline can be traced southwest and then west and northwest into the Moel Hebog Syncline; thence into the Snowdon Syncline; eastwards into the Crib Goch Syncline, and so back to the Arddu Syncline (figure 7). It has been suggested that this arcuate syncline is a rim structure which developed around a Caradocian magmatic dome, and was subsequently modified during the end-Silurian main deformation phase. The merits of the suggestion are considered later on pp. 39–44.

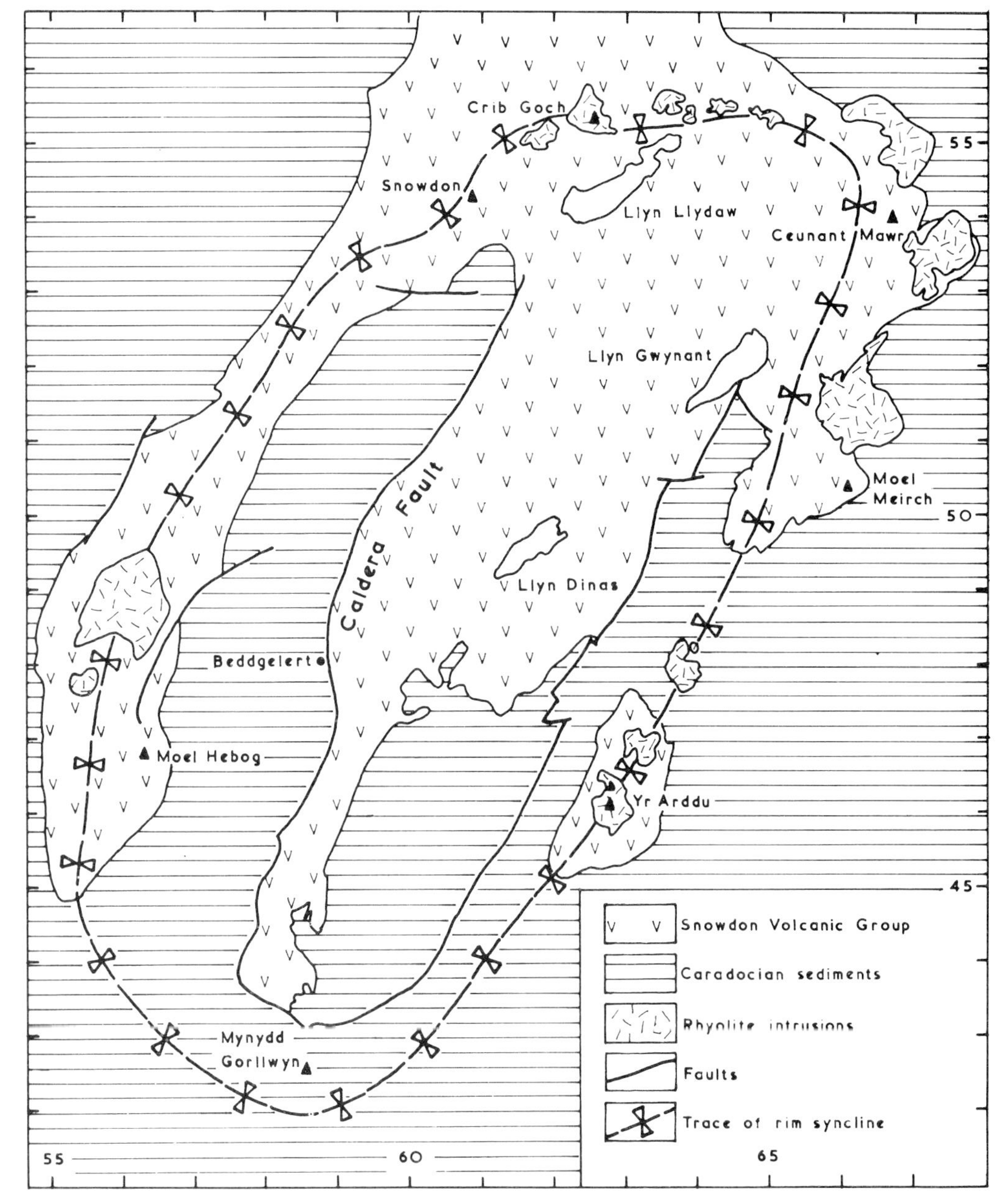

Figure 7. The postulated mid-Carodocian volcano-tetonic rim syncline of Bromley and Rast. Based on Bromley (1969) and Rast (1969).

1. The F_1 Structures

(i) *Introduction*

The outline geological map, figure 8 and, perhaps to an even better extent, the IGS 1:25 000 geological map, indicate that the prominent folds of the area are periclinal. This is particularly evident in the northern part of the IGS 1:25 000 map where, from west to east, the map emphasises the periclinal nature of the Cwm Idwal Syncline, the Cwm Tryfan Anticline and the Llynau Mymbyr Anticline. Such structures, which may be viewed in their entirety from suitable vantage points in the field, are here termed *major* structures. Figure 8 indicates that the major structures are themselves components of even larger-scale periclinal folds, which are here termed *maximal* structures. Such structures are too big to be viewed in their entirety in the field and are revealed only on small-scale geological maps. It is useful to distinguish as maximal structures the following.

1. The Arfon Anticline which runs through the Padarn Ridge from Llanllyfni in the southwest to Aber in the northeast. It has a southeastern limb common with the northwestern limb of the Snowdonia Syncline.
2. The Snowdonia Syncline (as opposed to the Snowdon Syncline which is merely a major structure), with a clearly defined southeastern limit against the northern flanks of the Harlech Dome.
3. The Ynyscynhaiarn–Cwm Pennant Anticline which trends more or less north–south and forms the western and southwestern margins of the Snowdonia Syncline.
4. The Llŷn Syncline which trends northeast–southwest and occupies most of Llŷn west of the Ynyscynhaiarn–Cwm Pennant Anticline.

(ii) *The Arfon Anticline*

The periclinal Arfon Anticline can be traced for over 45 km along its almost rectilinear axial trace. Internal deformation, as shown by the intense cleavage, has resulted in a tectonic thickening of the strata by a factor of about ×2, according to Wood (in Rast 1969). The strongly cleaved pelitic rocks from the Cambrian slate belt which runs from Nantlle to Bethesda, have provided the world's best roofing slates. The cleavage, S_1, is parallel to the axial surface of the major fold. Sandstones and conglomerates are sometimes converted to quartz–sericite phyllites or even schists. Exposure is rarely sufficiently adequate to allow one to distinguish component major folds, but the geometry of the folds which occur have been well described by Morris and Fearnsides (1926) from the Cambrian slate belt of the Nantlle area and, more recently, by Wood (1969) from Llyn Padarn. At Nantlle the folds are of similar type with vertical or steep northwesterly dipping axial surfaces. They are flattened buckle folds which are essentially the result of initial pure shear upon which later simple shearing has occurred. The northeastern closure of the maximal fold is well exposed in vertical section at Aber.

A prominent feature of the Arfon Anticline is the presence of strike faults. Their abundance is clearly evident in the geological maps of Morris and Fearnsides (1926), and it can be seen from their sections that whereas anticlinal hinges are a common feature, complete synclinal hinges are rarely preserved. It appears that the severe flattening was coupled with an essentially upward distension of the sedimentary pile, such that fold limbs are often extremely attenuated and commonly faulted, so leading to the development of faulted structures with incomplete hinges.

(iii) *The Snowdonia Syncline*

The Snowdonia Syncline is, on the whole, less intensely deformed internally than the Arfon Anticline and, although the rocks are strongly cleaved and pelitic formations have provided valuable roofing slates, phyllites are uncommon and schists very rare. Variable tectonic flattening has led to variable amounts of thickening of the strata and the few measurements made indicate thickening of about $\times\frac{5}{3}$. See, for example, Rast (1969) and Roberts and Siddans (1971).

Whereas the northwestern, southwestern and southeastern margins of the maximal syncline are reasonably clearly defined, the northeastern margin is particularly indistinct. The component major periclinal folds disappear eastward beneath the Silurian strata of the Denbigh Moors, having first, somewhere to the west of the Afon Conwy, swung in their trend from northeast–southwest to more or less east–west. In Arllechwedd, where the swing in trend occurs, the periclinal nature of the major folds is particularly obvious. Furthermore, folds tend to be more open and of greater wave length than those further to the southwest. To the southwest the wave length becomes less as the folds are constricted in the space between the northwestern flank of the Harlech Dome and the

Figure 8.
Structural map.

southeastern flank of the Arfon Anticline. The intensity of deformation is greater and folds may be overturned slightly towards the southeast on the northwestern limb of the Snowdonia Syncline, while in the south around Tremadog, several workers have claimed to have recognised important thrusts which presumably dip to the northeast and north.

In addition to the swing in trend from northeast–southwest in central Snowdonia to east–west in Arllechwedd, swings occur to eastnortheast–westsouthwest in the Dolwyddelan Syncline around the northern flank of the Harlech Dome, and to northnortheast–southsouthwest in the Moel Hebog Syncline along the flank of the Cwm Pennant Anticline. An interesting feature of the Dolwyddelan Syncline is that the northwestern limb is overturned towards the southsoutheast over a part of its length, despite the fact that it is on the southeastern limb of the maximal Snowdonia Syncline. Although no detailed measurements of thickness variations of individual beds on the major folds have been published, the main folds appear to have the geometry of flattened buckle folds.

The principal axial traces of periclines tend to run for distances of the order of 4–8 km, the longer traces being found in the areas of greater constriction in central Snowdonia.

Pelitic rocks and many tuffs usually have an axial-planar slaty cleavage, S_1, which shows the same swings in strike as the folds show swings in trend.

Strike faults are less common than in the Arfon Anticline, presumably because the flattening in these generally younger rocks is less, and the structures were generated at higher structural levels. Furthermore, they become less evident and less important with increased distance from the Arfon Anticline. We might expect the strike faults to re-appear on the southeastern limb of the Snowdonia Syncline, but such faults are not prominent on published maps. Instead, the 1:25 000 geological map shows a prominent thrust (the Penmorfa or Tremadog Thrust) at the base of the Carodocian sediments which has been folded with the sediments northwest of Tremadog. The thrust was first recognised by Fearnsides (1910) and confirmed by Shackleton (1959). Rather similar, but apparently non-folded thrusts were recognised by Fearnsides and Davies (1944) near Penrhyndeudraeth, but they regarded the thrusting as affecting a broad belt which it was not possible to represent on their map. It must be stated that there is little obvious evidence of thrusting in the field in any of these localities, but northwest of Tremadog, Shackleton (1959) has shown duplication of strata along low-angled, generally northeasterly dipping dislocations. An observer looking northwest from Minffordd cannot fail but to be impressed at once by the massive dolerite sills emplaced in the Glanrafon slates above the village of Tremadog and which dip at about 25° to the northeast. Turning to the northeast, he will be hardly less impressed by the thick acid sills within the Glanrafon slates at about the horizon of the Moelwyn Volcanic Group and dipping at similar angles but now to the northwest. The strike features formed by these massive horizons emphasise the southwestern closure of the maximal Snowdonia Syncline, and it is suggested here that significant bedding plane slip has occurred as a direct result of the presence of these competent tabular masses of igneous rock and that in the past the associated structures have been referred to as thrusts and slides.

In the area north of Tremadog the massive sandstone formations, such as the Prenteg Grits, possess an obvious axial-planar cleavage, S_1, whereas the Portreuddyn Slates beneath show instead a cleavage which has approximately the same strike as the formations but a rather steeper dip to the north. This latter cleavage can also be distinguished in some of the grits where it is seen to be cut by the more obvious S_1. The cleavage with the strike similar to that of the bedding is therefore designated S_0. It has been observed throughout this southwestern closure of the Snowdonia Syncline from Llyn Cwm Ystradllyn to Tremadog, and in the southeastern closure from Minffordd to Ffestiniog. It is thought to have been generated by flexural slip during an early stage of the development of the fold.

(iv) *The Cwm Pennant and Ynyscynhaiarn Anticlines*

The structural map (figure 8) shows that the Ynyscynhaiarn Anticline, with its northerly trend and plunge, can be regarded essentially as a major fold on the maximal structure of the Harlech Dome and as such should not be given the status of a maximal structure. However, it is also clear that the Cwm Pennant Anticline, with its northnortheasterly trend, can be regarded, together with the Ynyscynhaiarn Anticline, as the anticlinal structure that separates the Snowdonia Syncline from the Llŷn Syncline, and it is for this reason that it is given an equal status in this account.

The Ynyscynhaiarn Anticline, as depicted by Fearnsides (1910), has a slightly curved axial trace which trends overall just a few degrees west of north. It is asymmetrical with a steep western limb and a less steep northeastern limb. The map of Shackleton (1959) indicates that the axial trace swings through 60°, to a northeasterly trend, north of Golan, an aberration exceedingly difficult to interpret unless it represents an end-Tremadoc or immediately pre-Caradoc fold. This poorly exposed area is also of interest because offset from the Ynyscynhaiarn trace northwest of Golan is the southerly closure of the essentially periclinal Cwm Pennant Anticline. The periclinal structure, with plunge culminations and depressions, trends northnortheast and brings up in its core Ffestiniog slates and quartzites with perhaps some Arenig sandstones. Its limbs are bounded by strike faults and internal deformation is strong. The fold continues along the western side of Cwm Pennant for about 6 km and thereafter, a little offset to the east, is continued for about a further 2 km as a series of three small periclinal folds which bring up Arenig and Ffestiniog strata.

We thus have a series of anticlinal folds which swing through a gentle arc with trends to the northnorthwest in the south and swinging to almost northeast in the north. This arcuation is thought to be a primary feature. We shall return to this when we consider the question of the possible mid-Caradoc folds of the Snowdonia Syncline.

(v) *The Llŷn Syncline*

The Llŷn Syncline is a complex of folds within which it is possible to recognise two main areas for the purposes of description. These are (*a*) the Llwyd Mawr Syncline, and (*b*) the Llŷn Syncline *sensu stricto*.

(*a*) The periclinal Llwyd Mawr Syncline has been described by Roberts (1967) and it was here that the first attempt to systematise the polyphase nature of the late-Silurian to Devonian deformation was made. Exposure is good in the north and east but poor in the west and south, so that congruous minor folds are known in the northeast but hardly at all elsewhere. The principal axial traces of the component periclinal folds swing from a northeast–southwest trend in the north to about north–south in the south near Llanystumdwy, where Llandovery rocks are contained within the core of the major structure. West and south of Llanystumdwy the structure is unknown and can only be conjectured. The impression gained is that the Precambrian basement of Mona Complex rocks is as deeply concealed here as it is within the Snowdonia Syncline but, to the west and northwest, two factors begin to influence the structure revealed at the present level of erosion. The first of these is the complex of boss-like acid and intermediate intrusions emplaced in the country between Clynnog and Nefyn. If these are pre-F_1 intrusions then they may well be responsible—in part at least—for the swing towards east–west in the strike of the Cambrian slate belt from Llanllyfni to Clynnog, and perhaps also in part for the swing in strike from northeast–southwest towards north–south in the Llwyd Mawr Syncline, although this seems less likely. West of the Llwyd Mawr Syncline is the Llŷn Syncline proper, the western part of which seems to show the influence of the second factor, which is the now shallow depth of the relatively rigid Precambrian basement.

(*b*) The Llŷn Syncline is unfortunately not well exposed over much of this area, especially in its central and eastern parts where it enters Arfon and Eifionydd. It has a figure-of-eight-like trace on the ground. The western portion shows very well the influence of the close proximity of the basement, since here in the southwest the fold is broad, with the strike curving gently from east–west at Llanbedrog, through northwest–southeast at Botwnnog, to almost north–south near Cefnamwlch, and is interrupted by a series of radiating faults. It seems very likely that Llandovery rocks will occur beneath the drift in the neighbourhood of Rhyd-y-clafdy. The internal deformation in this southwestern closure is weak, but in the area of Madryn a dramatic change takes place: the strike of the beds abruptly turns through 90° to become eastnortheast–westsouthwest, dips are steep to vertical, and frequently overturned to the northwest. Such a fold limb may perhaps be the result of a response in the Ordovician cover to a steep reversed fault in the Precambrian basement.

In the south, at Llanbedrog, the southern limb also turns through 90° and the beds strike northeast towards Pwllheli, where another abrupt swing occurs and the strike becomes northnortheast as far as Four Crosses. Dips are steep to the northwest and it is here that a constriction in the syncline occurs which partially separates the structure into two areas. This separation is due, in part, to the Abererch Anticline. The fold plunges to the northnortheast and is probably faulted. Exposures in this eastern part of the Llŷn Syncline are very poor and much of the outline geological map is conjectural. Furthermore, the situation is made more difficult because the author has felt it necessary to re-interpret some of the information contained in the maps of Tremlett (1962, 1964, 1965).

An interesting feature of the southwestern closure of the Llŷn Syncline is that Crimes (1969a) has mapped a thrust at the base of the Tremadoc sediments and which is itself folded along with the overlying Ordovician sequence. Exposure is adequate in the immediate vicinity of Abersoch, but elsewhere along the course of the thrust exposure is virtually non-existent. The idea that north of the presumed thrust, which runs from Abersoch to beyond Botwnnog, the Ordovician sequence is to be regarded as allochthonous and as having moved into the present area from an unspecified distance away to the northeast is formidable; after all the sheet is pinned by a number of small boss-like intrusions such as those of Llanbedrog and Garn Fadrun.

The generally east and southeasterly dipping Arenig and Llanvirn strata around Aberdaron and Rhiw represent the western and northwestern limb of the Llŷn Syncline repeated by faults which downthrow to the west. Internal deformation is weak and reflects the influence of the less ductile basement which exists at just a shallow depth beneath the present surface.

2. The F_2 Structures

F_2 structures are rare and developed only on the mesoscopic scale; no large-scale F_2 structures have yet been identified. Typically, F_2 folds affect both bedding and S_1, and take the form of open, sideways-facing folds with rounded hinges and axial surfaces which are nearly horizontal. The trend is necessarily mainly Caledonoid. Folds are sometimes accompanied by the development of an axial-planar crenulation cleavage or, less commonly, a fracture cleavage. They have not been recognised in Arvonian or Cambrian strata, but only in Ordovician pelites and some cleaved tuffs. Roberts (1967) took this to suggest that they are some kind of incipient collapse structure developed as a response to the thickening of the sedimentary and volcanic pile which, in turn, resulted from the F_1 movements. F_2 folds and associated cleavage, S_2, have been described by Roberts (1967) from Llwyd Mawr and by Crimes (1969a) from western Llŷn. They may also be seen about Nant Ffrancon and again at Dolwyddelan, where folded cleavage planes were mentioned by Williams and Bulman (1930).

3. The F_3 Structures

The F_3 structures are much more widespread than the F_2 structures. They are again commonly developed on the mesoscopic scale, but larger-scale structures are occasionally present. Folds trend between westnorthwest–eastnortheast and north–south, and since they are usually developed in bedding and S_1 (F_2 structures being rare), plunges are commonly steep and to the northwest quadrant. Folds are usually accompanied by an axial-planar crenulation cleavage or fracture cleavage, S_3, which is sub-vertical in attitude and strikes roughly northwest–southeast. The folds are responsible for local swings in strike and have been described, for example, from Llwyd Mawr by Roberts (1967), from western Llŷn by Crimes (1969a), and from the area around Mynydd Mawr by Cattermole and Jones (1970). They occasionally become important as, for example, at Trefor in southwestern Arfon and near Conwy. Mesoscopic folds show rounded hinges and curved limbs, but occasionally more severe deformation in certain pelitic horizons leads to tighter folds with straight limbs and inter-limb angles of 50–60°. On other rare occasions the folds are brittle and show straight limbs with fractured hinges. Helm *et al* (1963) thought that the F_3 structures were largely responsible for major arcuations in the trend of F_1 fold-axial traces. The author is now of the opinion that whereas F_3 folds cause local swings in F_1 trends, they are not responsible for major arcuations. Rather, such swings are now thought to be largely primary features which were produced principally during the F_1 movements.

4. Faults

Some faults have already been described in connection with the F_1 movements. The strongly deformed Arvonian and Cambrian rocks, for example, were seen to be affected by a suite of steep strike faults which are an important feature in the Arfon Anticline and Cwm Pennant Anticline. They become less common at higher structural levels where deformation was less intense.

Several faults with the same trend in Arfon and Arllechwedd constitute very important structures. Thus the Dinorwic Fault, which more or less bounds the mainland along the Menai Straits, has a downthrow of at least

1000 m to the northwest. The latest important movements were of post-Carboniferous age and led to the preservation of the Lower and Upper Carboniferous rocks of Arfon. The Bangor Fault similarly has a downthrow to the northwest of about 300 m, and post-Ordovician movement has resulted in the repetition of Arenig, Lower Cambrian and Arvonian rocks. The Aber–Dinlle Fault in the northwestern limb of the Arfon Anticline again has a downthrow to the northwest of about 1000 m. Such important faults probably owe their initial development to Precambrian structures in the underlying Mona Complex, and Wood (1974) has suggested that they had an important influence on sediment accumulation and preservation during the Lower Palaeozoic. He pointed out that in Snowdonia the Arenig rests unconformably upon Ffestiniog Beds, but that between the Aber–Dinlle Fault and the Dinorwic Fault it rests unconformably on Lower Cambrian and that in northeastern Anglesey the Arenig has overstepped onto glaucophane schists of the Mona Complex. However, we cannot ascribe all such control to faults of this category since we have seen that an equally abrupt Arenig overstep occurs in western Llŷn in a direction almost parallel with the trend of these major faults.

Another feature of these early faults is that, just as on Anglesey (Greenly 1919), end-Silurian to Devonian movement on faults in the basement in Llŷn may have produced folds in the immediate cover. This is possibly the case in the overturned northwestern limb of the Llŷn Syncline.

The thrusts or slides of the southern and southwestern margins of the Snowdonia Syncline have also been mentioned and ascribed to movements akin to bedding-plane slip which took place at an early stage in the F_1 phase. The Abersoch Slide is more difficult to account for, however, and at this stage little more can be said.

A series of normal faults, more or less at right angles to the F_1 fold trends can be recognised. They are perhaps most clearly developed in the southwestern and northeastern parts of the Arfon Anticline and are clearly late products of the F_1 movements.

Various authors, including Rast (1961), Beavon (1963) and Evans (1968), have described faults which were active during the Ordovician, especially during the Caradocian, and which are supposed to have profoundly influenced the accumulation of sediments and volcanic rocks. Thus Evans (1968) described the Aber–Llanbedr Fault which runs southeastwards from Aber. To the north of this fault, over 3300 m of Glanrafon Beds accumulated, whereas south of the fault only 300 m were laid down. Rast and Bromley (1969) described a Caradocian fault system near Beddgelert, which runs from Beddgelert southwestwards to Moel Ddu, and then turns east to disappear beneath the alluvium of the Afon Glaslyn. It is claimed that here it turns northeast to become the Nanmor Valley Fault complex described by Beavon (1963), and that it then continues to Llyn Gwynant. It is further claimed that this fault represents the rim of a mid-Caradocian caldera which was filled with volcanics of one kind or another.

THE MID-ORDOVICIAN VOLCANIC RING STRUCTURE

The germ of the idea that a mid-Ordovician volcanic ring structure exists in Central Snowdonia was planted by Shackleton in a discussion which followed a paper by Beavon (1963) on the geology of an area east of Beddgelert. Beavon had suggested that a Caradocian dome had developed above a boss-like acid intrusion which had been emplaced near to the surface. This had caused the pre-existing volcanic and sedimentary rocks on the dome surface to slump off the flanks to give rise to the formation known as the Llyn Dinas Breccias. Shackleton suggested that the dome had also produced a rim syncline at its periphery as a direct result of the act of intrusion.

Subsequently, Rast (1969) and Bromley (1969) independently suggested the existence of a Caradocian volcanic ring structure in southwestern Snowdonia. They held that the dome had subsequently collapsed to form a caldera, the trace of which is seen today as the Beddgelert–Moel Ddu–Nanmor Fault systems; that the inclined sheet-like intrusions of dolerite, quartz–latite and rhyolite represent cone sheets; and that the course of the rim syncline is punctuated by a series of plug- and boss-like acid intrusions. The proponents have cited stratigraphical, structural and geophysical evidence in support of their claim, and their ideas are illustrated in figure 7.

1. The Stratigraphical Evidence

The Caradocian succession in Central Snowdonia can be summarised roughly as shown in table 4. The stratigraphical argument is concerned primarily with the Lower Rhyolitic Group.

Table 4. The Caradocian succession of central Snowdonia.

Group	Lithology
Nod Glas Group	Black Shales
Snowdon Volcanic Group	Upper Rhyolitic Group Basic Group (≡ Bedded Pyroclastic Group) Lower Rhyolitic Group { Lower Rhyolite Tuff; Llyn Dinas Breccias; Pitt's Head Ignimbrite }
Glanrafon Group	Volcanic sands, slates, etc, with andesitic and rhyolitic volcanics, such as the Glôg and Moelwyn Volcanic Groups, and the Capel Curig Volcanic Group

The Glanrafon Group is essentially marine and fossiliferous. An unconformity occurs at the base of the Pitt's Head ignimbrites and was said to be greatest around Snowdon, for in this vicinity only one component ash flow has been mapped, whereas to the northeast and southwest two flows have been recorded. The change from marine Glanrafon Beds to subaerial Pitt's Head ignimbrites, with maximum uplift near Snowdon, was regarded as evidence for pre-Pitt's Head magmatic doming centred near the present Llyn Dinas. It may be illustrated as shown in figure 9.

There is a further unconformity beneath the succeeding formation, the Llyn Dinas Breccia. It is locally very strong; indeed, it cuts out over 1500 m of Glanrafon Beds at Aberglaslyn and it is visibly very strong at Pont-y-gromlech in the Pass of Llanberis. This was regarded as further evidence for magmatic doming. The Llyn Dinas Breccia formation itself was regarded as a lahar, the product of a volcanic mud flow.

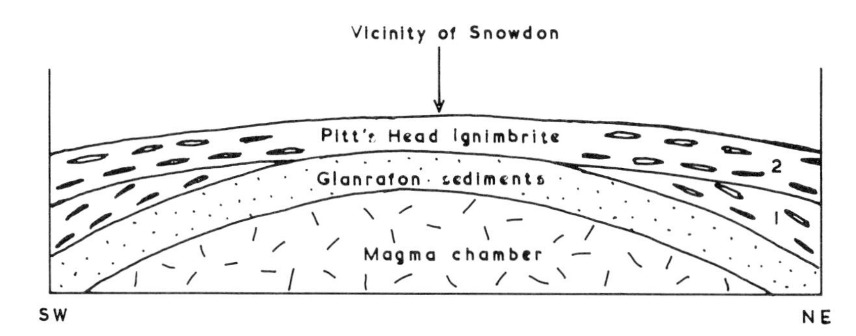

Figure 9. Diagrammatic section to illustrate the postulated pre-Pitt's Head magmatic dome of Bromley and Rast.

There is yet another unconformity beneath the Lower Rhyolitic Tuff formation. It is strongest in Central Snowdonia where the tuff is both thickest and mainly subaerial. It passes into fossiliferous marine tuffs to the north, east and southwest. This was regarded as additional evidence for magmatic doming in south–central Snowdonia.

The Lower Rhyolitic Tuff passes upward into marine, fossiliferous tuffs and is overlain by the Basic Group which, again, is marine and fossiliferous. The marine incursion was regarded as evidence for volcanic collapse and caldera formation consequent upon eruption. In other words, the evacuation of the magma chamber led to caldera formation and the subsequent marine incursion.

It is essential to the argument for the dome that the Pitt's Head ignimbrites should not be equivalent to the Llwyd Mawr ignimbrites away to the west, although this correlation has been suggested by Shackleton (1959) and Roberts (1967). This is because on Llwyd Mawr, to the west of the dome, the Llwyd Mawr ignimbrites rest unconformably upon Llanvirn slates, all the intervening Glanrafon Beds having been cut out. Geochemical evidence, however, supports the correlation. G Hendry (1973 personal communication) writes '. . . the Llwyd Mawr ignimbrite is closer in its chemistry to the Pitt's Head flows than any other unit in Snowdonia'. The probability therefore is that they are equivalent and so the Pitt's Head unconformity is greatest not around Snowdon, but well to the west of the postulated ring structure on Llwyd Mawr. That the sub-Pitt's Head unconformity strengthens westwards can be seen from the IGS 1:25 000 geological map, and on the western limb of the Moel Hebog Syncline the Glanrafon Beds have been reduced to a thickness of about 180 m.

Secondly, if the formation of the rim syncline is a result of dome formation, then eruptions from the caldera should have flooded the rim syncline around the dome and the volcanic formations should thicken into the syncline. There is no evidence that this has happened.

However, there is clearly an unconformity beneath the Lower Rhyolitic Tuff, and there is certainly lateral passage into fossiliferous marine sediments away from Snowdon. The raw thickness variations are 360–450 m on Snowdon; 55 m on Moel Hebog; and 47 m at Roman Bridge, Dolwyddelan. If allowance is made for tectonic thickening, then a better approximation is 270 m on Snowdon, to 33 m on Moel Hebog, and to 30 m at Dolwyddelan. Unrolling the folds (which can only be an approximation because they are flattened buckle folds) suggests an edifice rising perhaps some 270 m above the sea which was some 9 km away to the southwest from the highest point. If the Lower Rhyolitic Tuff had been laid on a planar, horizontal surface, this gives a slope of about 2° to the upper surface; but whatever the nature of the sub-Lower Rhyolitic Tuff surface, a low-profile eruptive centre near Snowdon is indicated. That it should be of low profile is demanded by the ignimbrite units within the formation which retain a reasonably uniform thickness over considerable areas and were clearly not deposited on a significant slope. Howels *et al* (1973) reached the same conclusion as a result of a more sophisticated study of the Lower Rhyolitic Tuff in eastern Snowdonia. It seems unlikely, therefore, that doming on the scale required by the hypothesis, occurred.

2. The Structural Evidence

Three lines of evidence have been described in support of the hypothesis.

(i) A syncline, with plunge depressions and culminations can be traced in an oval, the axes of which measure 12 km northeast–southwest by 8 km northwest–southeast. This was interpreted as a rim syncline which formed at the margin of the dome as a result of the emplacement of the magma blister. It passes through Moel Hebog, Mynydd Gorllwyn, Yr Arddu, Moel Meirch, Ceunant Mawr, Crib Goch and Snowdon.

(ii) Rast (1969) recognised a fault, the Beddgelert fault which, it is argued, is of Ordovician age. Beavon (1963) recognised a similar fault east of Beddgelert, the Nanmor Fault. They are thought to be connected in the south beneath the alluvium-covered ground of the Afon Glaslyn, and are interpreted as constituting the ring fault which marks the caldera wall.

(iii) The end-Silurian S_1 is a slaty cleavage in pelitic rocks and has a generally Caledonoid trend. This cleavage is broadly axial-planar to parts of the rim syncline, such as the component Moel Hebog and Arddu Synclines, but elsewhere, particularly in the north and south, it cuts across the trace of the rim syncline at high angles. This is taken to suggest that the rim syncline was in existence before the cleavage was generated.

If the course of the rim syncline is considered, it is true that in the southwest a periclinal syncline with a long axis trending westnorthwest–eastsoutheast runs through Mynydd Gorllwyn. If the map of Shackleton (1959) is examined, however, several more fold axial traces can be recognised, which run generally parallel to the curving strike of the Prenteg Grits (figure 10). Furthermore, it is possible to continue the axial trace of the Bryn Banog Anticline to the westsouthwest so that it separates the Moel Ddu Syncline from the Mynydd Gorllwyn Syncline. Clearly, whatever was responsible for the curvature of the Bryn Banog axial trace was also responsible for the other axial traces curving away in parallel fashion from Mynydd Gorllwyn to Moel Hebog. It is very difficult to see how this could have been brought about by a rising magmatic dome.

On the other hand, it is not difficult to relate these structures to the maximal Snowdonia Syncline. That movements of a bedding-plane slip type occurred in this closure of the maximal periclinal syncline is evident from such observations as the presence of a cleavage within the Portreuddyn Slates and the pelitic members of the Prenteg Grit formation which has the same curving strike as the grits. This cleavage, S_0, the strike of which follows the strike of the bedding, usually dips northward more steeply than the bedding. Its wide distribution indicates that it is of tectonic rather than volcano-tectonic origin; it is therefore almost certainly end-Silurian to Devonian in age, and is due to flexural slip. As

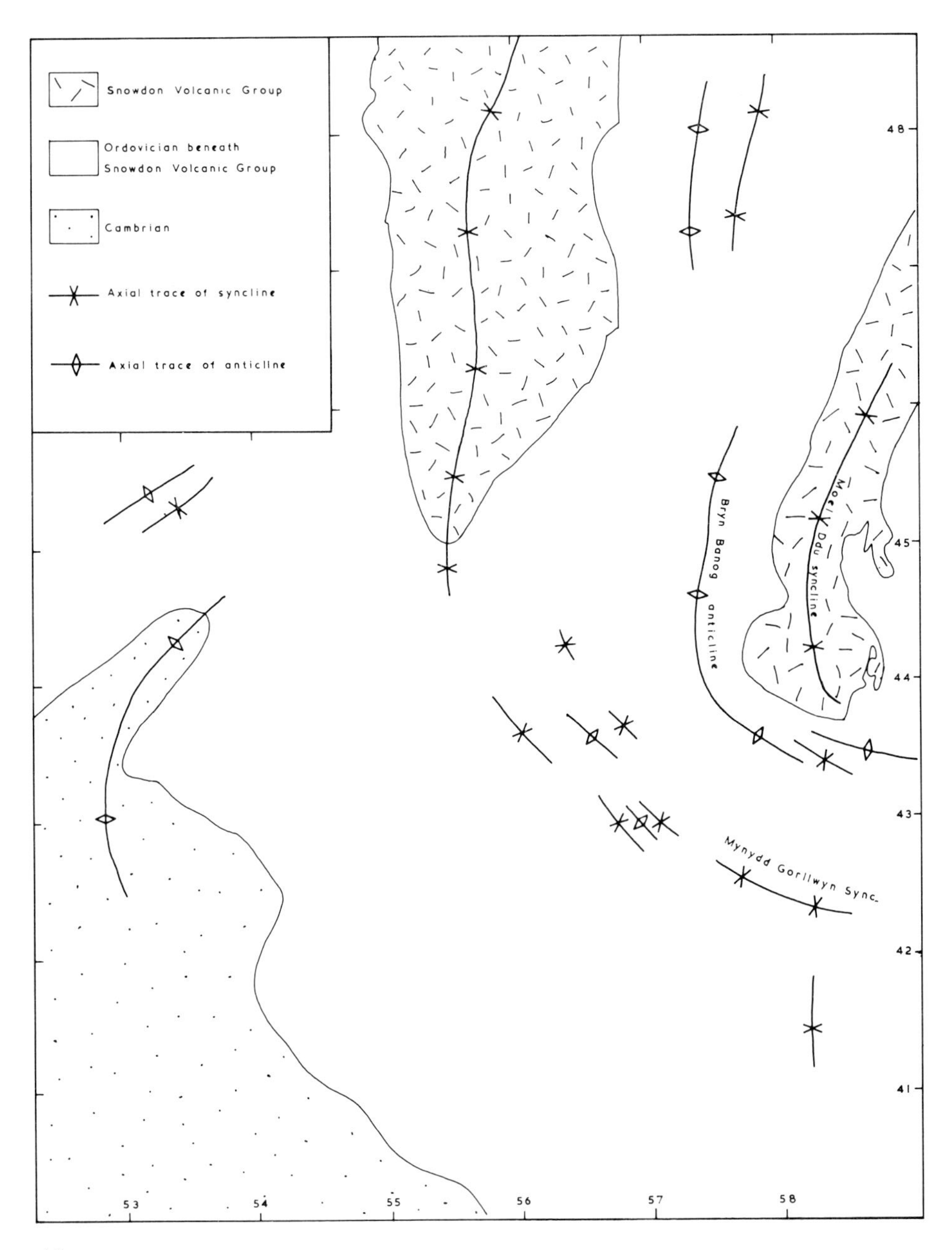

Figure 10. Fold-axial traces in the southwestern part of the postulated mid-Carodocian volcanic ring structure.

the Mynydd Gorllwyn Syncline is crossed the relationship between S_0 and bedding is maintained and they have been folded together. The Mynydd Gorllwyn Syncline is therefore of end-Silurian to Devonian age, rather than mid-Ordovician.

The dislocations, however, are equally important. Shackleton (1959) recorded northeasterly inclined thrusts at Pen Morfa; Mynydd Gorllwyn itself is cut by two faults inclined to the northnorthwest and on which dextral slip appears to have taken place; and the faults which chop up the outcrop of the Glog volcanics at Llyn Cwm-bach may be small dextral tears. It is suggested here that the overall northwesterly trending swarm of folds and the accompanying dislocations are due to movements of a bedding-plane slip type in the constricted southwestern closure of the maximal Snowdonian Syncline.

The trace of the proposed rim syncline then disappears beneath the alluvium of the Afon Glaslyn and, in view of the extent of the drift cover, it is largely conjecture to suggest that the Mynydd Gorllwyn Pericline continues through the unexposed ground to become the Arddu Syncline.

The Arddu Syncline forms the eastern part of the rim syncline. It trends southwest–northeast over much of its length until, at Moel Meirch, it swings towards eastnortheast and dies out. A plunge culmination then separates it from the Dolwyddelan Syncline, which trends away towards the eastnortheast. North of Moel Meirch is the intrusive rhyolite of Cerrig Cochion, and the next recognisable fold north of this is the Llyn Gwynant Syncline which, in the course of 2 km, swings from a northeast to an eastnortheast trend before dying out. It has been suggested, however, that the Arddu Syncline continues through this area occupied by the intrusion of Cerrig Cochion and the Llyn Gwynant Syncline, and then curves northwest around Crib Goch. There is very little evidence on the IGS 1 : 25 000 geological map to substantiate this. It is further suggested that the syncline continues from Crib Goch to Snowdon.

It will be appreciated that in the southwest the rim syncline was said to exist in the sediments beneath the volcanics as at Mynydd Gorllwyn and Yr Arddu; but in the northeast it is said to be present high in the volcanic sequence which accumulated after the dome and its attendant rim syncline is thought to have formed. This is clearly an impossibility since the rim syncline, which is stated to have been formed in the Glanrafon Beds by the uprise of the magmatic dome, would by this later time have become filled with volcanic ejectamenta, particularly since in this vicinity the Lower Rhyolitic Tuff is at its thickest and consists largely of ignimbrites. The one place where we should be unable to detect any syncline today is indeed in the volcanic sequence in this part of the postulated structure. There is no doubt, however, that the southwest closure of the large Glyder Fach–Cwm Tryfan Anticline is slightly anomalous, but this is probably due to the influence of the thick wedge of the Glyder Fach Breccia formation.

3. The Distribution of Intrusive Igneous Rocks

Bromley (1969) argued that the dolerite intrusions occur as dykes well below the volcanics; as sills at the base and within the volcanics; and that within the area enclosed by the rim syncline, dolerite intrusions are absent.

It was pointed out that acid sheets were emplaced only on the western limb of the Snowdon Syncline, the eastern limb of the Arddu Syncline, and the northern limb of the Crib Goch Syncline. Composite sheets occur on the southwestern limb of the Arddu Syncline; their margins are dolerite and the interiors are of quartz–latite. The dolerites, acid sheets and composite sheets were all interpreted as a cone sheet complex surrounding the central dome.

It was emphasised that high-level rhyolite plugs were usually emplaced in the axial region of the rim syncline. The presence and arrangement of these suites of intrusive rocks suggested that the caldera was underlain by a plug of acid rock overlying a cylindrical plug of basic rock. The author's understanding of the Bromley–Rast hypothesis is illustrated diagrammatically in figure 11.

The IGS 1 : 25 000 geological map reveals that dolerites are not absent from within the area of the proposed rim syncline. The sheet-like dolerite intrusions outside the axial trace of the proposed syncline were thought to represent cone sheets, but if we look beyond the rim syncline it can be seen that the same dolerite sheets extend across Gwynedd, from the west of Llŷn to Conwy, as a folded, discontinuous series of sills.

The acid sheets on the western and eastern sides of the proposed structure were also thought to represent cone sheets. G Hendry (1973

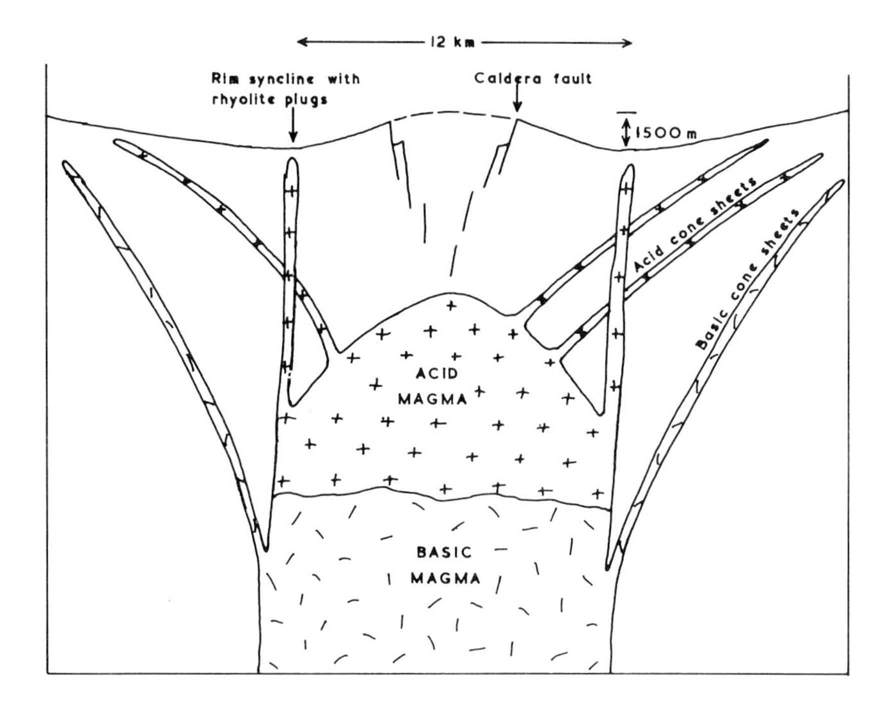

Figure 11. Diagrammatic section to illustrate the author's understanding of the Bromley–Rast hypothesis.

personal communication) states that they show significant geochemical differences, and it is very unlikely that they can be equated. They are not strictly coeval and are therefore unlikely to be cone sheets.

High-level acid plugs were thought to be confined to the rim syncline, although they also occur extensively outside it. Furthermore, the large Castell Rhyolite in the Moel Hebog Syncline is shown in Rast (1969) as a plug, whereas it is a sill-like sheet.

4. Geophysical Evidence

It has been pointed out that the rim syncline coincides with gravity highs (Rast 1969). This was interpreted as the result of the presence of a plug of basic rock with an overlying granite cap beneath the caldera. The basic magma was thought to have melted the crust to produce the acid magma and hence the acid eruptive rocks of the Snowdon Volcanic Group.

The map of Griffiths and Gibb (1965), however, does not show that the rim syncline coincides with gravity highs, although two anomalies do lie just west and northeast of the axial trace of the Moel Hebog Syncline. If the proposed ring structure is underlain by a basic plug, as has been suggested, one is entitled to wonder why the high is not sited more centrally beneath the total structure?

Finally, there is some experimental evidence that bears on the hypothesis. Ramberg (1970) has shown, using scaled centrifuged models, that a rim syncline can only develop around an intrusive plug when there is minimal viscosity contrast between the intruding mass and the country rock. The country rock was, of course, solid, but the postulated intruding mass was liquid because the hypothesis requires it ultimately to have broken through to the surface to give ignimbrites and other volcanics, as well as near-surface plugs, cone sheets, etc. A rim syncline cannot be formed by a Newtonian fluid but it must be admitted that we know very little about the viscosities of magmas.

5. Conclusion

Scrutiny of the evidence fails to support the suggestion that a mid-Ordovician volcano-tectonic ring structure of the type envisaged by Rast and Bromley exists in the Snowdon area. Nevertheless, the hypothesis has served as a stimulus to workers in the area and has focused attention on the search for volcanic centres and the relations between these and sedimentation.

Part II

Field Guide

Introduction

It is not anticipated that any one person would work his way methodically through the entire 31 excursions in one stint. Such an undertaking would need the better part of two months if due allowance is made for the weather usually experienced in North Wales. Rather, it is hoped that the user, having read the outline of the geology of the area in Part I, will be in a position to select excursions which contain those aspects of the geology which interest him most. Alternatively, if a reader wishes to examine the geology of an area in which he happens to be staying, say Porthmadog, he has merely to follow Excursion 8; if he wishes to know something of the geology of a group of hills over which he intends walking, say the Carneddau, he would use Excursion 26, and so on. Nevertheless, for undergraduate students of geology, time always seems to be at a premium and so the following programme is suggested for about one week's work.

1. Southwestern Llŷn

Excursion 1, Locality 7: Trwyn Maen Melyn to Parwyd.
Gwna mélange and Gwyddel beds of Mona Complex; increase in metamorphic grade eastward to schists.
Locality 9: Meillionydd.
Foliated and unfoliated gneisses and migmatites of the Mona Complex.
Excursion 2, Locality 1: Mynydd Penarfynydd.
Llanvirn (*bifidus* zone) slates. Layered mafic–ultramafic sill of Ordovician age.
Excursion 3, Locality 1: Trwyn Carreg-y-tir.
Hell's Mouth Grits and Mulfran Beds (Lower Cambrian turbidites).
Locality 2: Trwyn Cilan.
Cilan Grits (Middle Cambrian turbidites).
Locality 3: Trwyn Llech-y-doll
Arenig conglomerate and sandstones resting unconformably upon Middle Cambrian Caered Mudstones. Abundant trace fossils in Arenig.

2. Northern Arfon

Excursion 11, Locality 2: Gored-y-gut.
Cambrian sandstones and conglomerates overlain by fossiliferous Lower Carboniferous shales, sandstones and limestones.
Locality 5: Llanfair-is-gaer.
Upper Carboniferous conglomerates and marls cut by mafic Tertiary dykes.
Locality 7: Afon Seiont.
Classic section in graptolite-bearing Arenig and Llanvirn Slates.
Excursion 12, Locality 1: Clogwyn Melyn.
Rhyolitic ignimbrites of Arvonian Series close to the axial trace of the Arfon Anticline.
Locality 2: Mynydd Cilgwyn (but omitting initial exposures in the two small quarries).
Intensely deformed (flattened) Glôg Grit and Cilgwyn Conglomerate.
Locality 4: Alexandra–Moel Tryfan slate quarry. Sequence and tectonics in the Cambrian Slate Group, in the southeastern limb of the Arfon Anticline.

3. Southern Arfon

Excursion 14: Moel Hebog.

This compact area includes a complete succession through the Snowdon Volcanic Group together with underlying Caradocian sandstones and slates. Intrusive rocks include acid and basic types. The Moel Hebog Syncline, a component in the Snowdonia Syncline, is traversed.

4. Southern Arllechwedd

Excursion 27: Capel Curig to Trefriw.

Rocks range from Caradoc to Ashgillian, and include the Crafnant Volcanic Group and the Trefriw Tuffs. A volcanic vent is seen at Capel Curig cutting fossiliferous Caradoc sandstones. Sediments above the Crafnant Volcanic Group include fossiliferous Caradoc slates of the *clingani* zone and Ashgillian slates with diplograptids. Major folds within the maximal Snowdonia Syncline are traversed.

Llŷn

EXCURSION 1. THE PRECAMBRIAN OF WESTERN LLŶN

Particular attention is paid to the problem of the relationship between the gneisses and the low-grade rocks because it is difficult to over-emphasise the importance of this relationship to interpretations of the geology of North Wales.

There is probably at least two days' work involved in examining the exposures in the following localities. If only one day is available however, the best section would be Locality 7 (Trwyn Maen Melyn to Parwyd), followed by Locality 8 (the Meillionydd migmatitic gneisses), and finally Locality 3 (the Sarn Granite). In this series of exposures one sees the Gwyddel and Gwna Groups in their lowest metamorphic state, the rapid increase in grade to the east which converts them to schists, the gneisses to the east of the schists and finally, further east again, the migmatitic gneisses and the homogeneous Sarn Adamellite.

Transport is essential for travel between exposures.
Use the excursion geological map, figure 12.
References: Matley (1928), Matley and Smith (1936) and Shackleton (1956).
1:50 000 OS sheet No. 123; or 1:25 000 sheets Nos SH12, 22, 23 and 34.

Locality 1: Penrhyn Nefyn [296410]

If the visit is made in the holiday season, cars are best left in a car park 600 m along the road from Nefyn to Morfa Nefyn. Out of season, cars can be left at the bottom of the steep road leading down to the beach at Porth Nefyn.

Exposures of Precambrian rock begin beneath the wall immediately north of the huts. Low tide is helpful but not essential.

The cliffs are of drift resting on a low wave-cut platform of Precambrian. The first exposures encountered in a south to north traverse of the eastern side of the point are felsic gneisses. These are massively banded, the bands ranging from 30 cm to 1 m in thickness; the banding strikes at 056° and dips are from 75° northwest to vertical. The gneisses are cleaved and sheared, and the cleavage dips 45° towards 308°. Mafic gneisses (amphibolites) with acid veins structurally overlie the felsic gneisses. The mafic gneisses are migmatitic, the felsic veins being sometimes parallel with the foliation, sometimes discordant, and sometimes apparently ductless bodies. At a distance of about 30 m north of the wall by the huts, the amphibolites, with more felsic veins, give way to greenschists via a narrow zone in which pods of coarse-grained mafic gneiss are enclosed in greenschist. This has been interpreted as a fault by Matley (1928) and as a gradational boundary by Shackleton (1956).

The greenschists, in which cleavage dips 45° towards 326°, have been derived from pillow lavas, and partly deformed pillows and pods of chert can still be distinguished. The flattening which produced a flow cleavage in the schists has produced a fracture cleavage in the chert pods at right angles to the cleavage in the schists.

5 m south of a small sandy bay the greenschists are cut by a 10 cm thick mafic dyke. This Tertiary dyke trends at 310° and can be seen to be stepped as well as discontinuous as it is followed northwestward.

On the northern side of the small bay the greenschists are derived from laminated tuffs and cherts, and pass up into more homogeneous quartz–albite–muscovite–chlorite schists. The schists are cut by a set

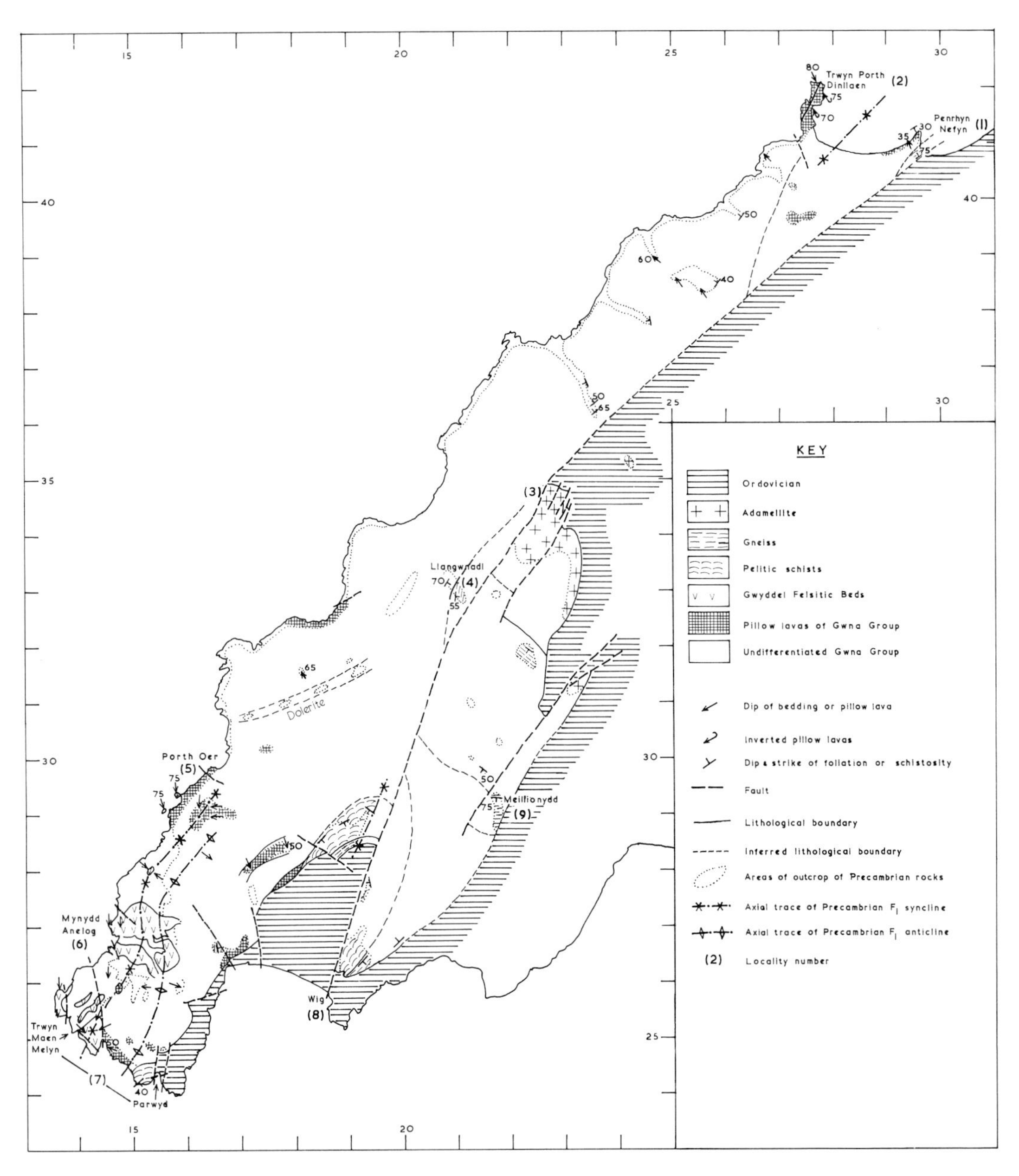

Figure 12. Excursion 1. The Precambrian of western Llŷn. Largely after Shackleton (1956), with permission.

of prominent, vertical quartz–chlorite veins trending at 315°, and are affected by two sets of Precambrian minor folds: one set, F_1, plunging at 10° towards 066°; and a second set, F_2, associated with a fracture cleavage, S_2, striking at 314° and dipping 80° northeast. Greenschists with fine cherty bands continue to the point.

At the point a vesicular dyke cuts the greenschists and trends at 312°. The greenschists now dip 50° towards 010°. Some 25 m around the point, on the western side, are prominent orange-brown-weathering carbonate–quartz veins. The carbonate carries bornite and chalcopyrite, while quartz in the centres of the veins carries arsenopyrite. The veins trend at 300° and dip 60° southwest.

Locality 2: Trwyn Porth Dinllaen [2741]

Cars may be left in Morfa Nefyn. The road along the promontory beyond the Golf Club House is normally closed to cars.

The promontory consists of pillow lavas of the Gwna Group. The features of pillow lavas can be examined anywhere along the shore. Exposures are magnificent. Dips are steep to the northwest and, as Shackleton has pointed out, the convex upper sufaces and concave lower surfaces of individual pillows indicate that the lavas are overturned and young to the southeast.

Locality 3: Mountain Cottage Quarry [230347]

There is no car parking problem. At a point just southeast of the quarry, steps have been placed in the roadside wall and give access to the largely overgrown quarry. The best exposures are in a small subsidiary quarry in the northwestern part at the top of the main quarry. **The exposure should not be hammered.**

Cleaved Arenig shales, siltstones and fine-grained sandstones overlie a 10 cm thick basal conglomerate which, in turn, rests unconformably upon 'granite'. The surface on which the Arenig sediments rest is uneven and cut by a few small reversed faults, each with a throw of a few centimetres. *Bolopora undosa* and horny brachiopods have been obtained from this exposure.

The 'granite', which is best examined in the lower quarry, is in fact an adamellite. It is cut by closely spaced joints along which it usually breaks when hammered. Prior to the work of Matley and Smith (1936), majority opinion was that the adamellite was intrusive into Ordovician sediments, but it is now seen as part of the Mona Complex.

Locality 4: The Llangwnadl River Section [208332]

A car can be left at St Gwnadl's Church. The section can be entered from the road, a few metres upstream from the church. The traverse begins in low-grade schists and ends in high-grade, migmatitic amphibolites.

Some 25 m upstream from the church the stream is interrupted by a small weir. Just below the weir a quartz–albite–muscovite–chlorite schist is exposed in the right bank; the foliation dips 70° towards 102°. About 10 m upstream from the weir a greenstone, which may represent a meta-pillow lava, is exposed. A further 10 m upstream, a pelitic greenschist with quartzo-feldspathic, ductless segregations is seen structurally overlying a poorly schistose metabasic rock.

Upstream, the stream makes an abrupt right-angled bend. Downstream from the bend, greenschists and amphibolites are exposed. Some are severely crushed and others less so. The most prominent exposure is cut by a non-schistose, fresh, black, basic rock—presumably a Tertiary dolerite dyke? 30 m upstream from the prominent bend, massive basic gneisses with large prisms of hornblende crop out. Some show irregular bands and knots of quartzo-feldspathic segregations. The foliations strike between 335° and 010° and dips are vertical to 65° east. These migmatic gneisses are exposed upstream as far as a small waterfall near an old sluice. Thereafter, there are no further exposures to the bridge and the B4417 road.

Locality 5: Porth Oer [165298]

Exposures on the southwestern side of the bay provide a good place to study minor structures in the Mona Complex. In the holiday season, cars can be left in the field near the end of the road which is given over to a car park.

The first exposures encountered are finely banded, green, khaki and brown basic tuffs with occasional carbonate-rich laminae, dipping 70° towards 200°. The earliest Precambrian cleavage is a slaty cleavage, striking at about 60° and dipping steeply northwest. The bedding and first cleavage are folded about a set of steeply plunging, minor fold axes (F_2 in this locality) with an associated axial-planar strain-slip cleavage. The F_2 minor folds plunge at 70° towards 185°; the strain-slip cleavage strikes north–south and dips at 80° west. The exposure is much chopped up by a set of very small faults.

Some 20 m further seaward, and structurally lower, can be seen green tuffs with isolated pillows followed by red tuffs. First minor folds, F_1, are now obvious and possess an axial-planar, slaty cleavage. Folds plunge at 10° towards 240° and the slaty cleavage, S_1, dips 70° towards 328°. Seaward, the tuffs are underlain by pillow lavas.

Locality 6: Mynydd Anelog [153273]

The junction between the Gwna mélange and the overlying Gwyddel Beds provides the main interest. The approach is along a narrow lane. There is just sufficient space for two cars to be parked clear of the road at the junction at [156275]. Take the steep lane leading southwest towards the summit of Mynydd Anelog.

Southeast of the summit there are small exposures of white-weathering, finely laminated, porcellanous, vitric tuff of the Gwyddel Group. The bedding strikes at 070° and dips at about 60° southeast. At the summit the strike of the beds has swung to 100° and the dip is 70° south. There is prominent development of minor F_2 folds which plunge at about 50° towards 210°. A traverse northward from the summit takes one down the succession from the acid tuffs of the Gwyddel Group through an intervening pillow lava into laminated basic tuffs and greywackes, sometimes graded, of the Gwna Group. Presumably, all the rocks accumulated in a submarine environment.

Figure 13. Cliff top at Mynydd Anelog [147272]. Bedded rhyolitic tuffs of the Gwyddel Felsitic Beds rest on pillow lavas of the Gwna Group. A southeasterly verging F_1 minor fold-pair with an associated axial-planar cleavage (S_1) is developed.

The junction can be followed from just north of the summit southwestward to reach the excellent exposures at the cliff top about 100 m southwest of a prominent inlet and wire fence. It can be seen that the Gwyddel Beds overlie the Gwna Group which is here in the form of a mélange (figure 13).

At the junction, the Gwna Group is represented by a red, basic tuff with isolated pillows and large clasts of dolomite. The contrast with the uniformly bedded Gwyddel Group is striking. Again two sets of folds are developed: F_1 minor folds, with an associated axial-planar slaty cleavage, plunge at 45° towards 142°; and F_2 minor folds plunge at 50° towards 282°. The anomalous trend of the F_1 structures is presumably partly due to refolding about F_2 axes, but F_2 structures too are anomalous, so we may presume that both have been swung by a subsequent set of structures which have not been distinguished by the author.

Locality 7: Traverse from Trwyn Maen Melyn [138252] to Parwyd [155243]

There is ample parking space in the car park just inside the National Trust area.

The main interests are firstly the spectacular exposures of the Gwna mélange and, secondly, the rapid passage from unmetamorphosed and low-grade Gwna Group in the west into medium-grade schists in the east.

Walk northward along the cliff from Trwyn Maen Melyn for 250 m over exposures of Gwna mélange. The mélange has clasts of dolomite, limestone, quartzite, phyllite, greywackes and pillow lavas in a green, basic matrix. Some of the blocks can be seen to measure several metres across, but it is not until you reach a point [137253] about 250 m from Trwyn and look back southward at the cliffs that the huge size of some of

Figure 14. Cliff at Trwyn Maen Melyn. The Gwna mélange includes blocks of white-weathering quartzite, the largest of which is 17 × 25 m.

the blocks becomes apparent (figure 14). Shackleton (1956) records that the lenses of pillow lavas and greywackes in the mélange remain the right way up, and this can be verified in the field by the shapes of the individual pillows and by grading in the greywackes.

Retrace your path along the cliff to St Mary's Well and walk over the summit of Mynydd Gwyddel. The laminated, white-weathering, porcellanous, acid tuffs of the Gwyddel Group can be seen to occupy a syncline.

At Porth Felen a pair of faults trending north–south brings up the Gwna Group. Between the faults the Gwna Group is represented by cleaved, well bedded, red and green, basic tuffs. On the eastern side of the easternmost fault are red and green pillow lavas which become tuff-bearing and very strongly cleaved 100 m southeast of the fault. After a further 50 m thin, grey slates with greywackes dip 50° towards 288°. The horizon is underlain by mélange containing lenses and blocks of dolomite, cherty dolomite, and pillow lavas with jaspers. After 30 m of this mélange, the succession returns to pillow lavas which face upward. Then comes a further obvious mélange much as before, and this is again followed by pillow lavas. Deformation seems to increase to the southeast and several sets of minor folds can be seen in the schists and phyllites, including a set of sideways-facing folds, most plunging at various angles towards about 210°. Rock types include quartz–albite–muscovite–chlorite schists, muscovite–dolomite–quartz schists and chlorite–epidote schists.

On Trwyn Bichestyn, opposite Carreg Ddu, it is possible to determine that the schists represent the metamorphosed Gwna mélange because it is possible to distinguish basic schists, calcareous schists and marbles, along with included relict pillows and masses of chert. The rocks are cut by post-metamorphism faults and deformed by kink bands. The minor folding is not easy to sort out. Further east at [153242] overlooking Parwyd, the rocks can be clearly seen to be meta-pillow lavas.

From this point, Parwyd can be seen to be formed as a result of a pair of faults letting down Arenig sediments. The cliff top can be followed around the head of the bay in National Trust ground. If care is taken, it is possible to descend the grassy cliff where mafic gneiss can be seen with a thin pocket of Ordovician sediments resting nonconformably upon them at [155243]. The gneisses are mainly banded amphibolites, some parts of which are garnet-bearing. They occupy a strip some 30 m wide and 300 m long. The foliation strikes at about 160° and dips at about 50° to the east. The best exposure of the Ordovician unconformity is difficult to reach—great care, a rope and the ability to use it are necessary. Matley (1928) records some 60 cm of grit or sandstone overlain by shales.

Back at the cliff top, a detour of a few metres to the east brings you to the outcrop of the Pen-y-cil dolerite sill. The sill is about 100 m thick and is emplaced into *D. extensus* zone sediments of the Arenig. The rock is spilitic in mineralogy, the plagioclase having altered to albite plus pumpellyite. Prehnite is rare, but the rock probably belongs to the prehnite–pumpellyite facies of very low grade regional metamorphism.

Locality 8: Wig [187258]

From Aberdaron walk 1200 m east along the beach to the cliff exposures at Wig.

The sandy drift forming the cliff gives way to drift with angular blocks of Ordovician sediments and then into crudely bedded, blocky drift banked against Ordovician sediments which, in turn, are faulted against Precambrian schists.

The Ordovician sediments consist of alternations of pelite, sandstone and granule conglomerate. The rocks are folded into an anticline and complementary syncline plunging at 48° towards 178°. The cleavage, which is well seen in the anticline, fans slightly round the hinge and in the common limb has a strike of 350° and dips 70° west. The silts and pelites are strongly bioturbated and belong to the *D. extensus* zone of the Arenig.

The schists (regarded by Matley as belonging to the Penmynydd zone of metamorphism) can be seen to be derived from the Gwna mélange because within them flattened, ellipsoidal lumps of pink and white quartzite can be distinguished. The schistosity has an overall strike of 010° and dips 80° west. Around the small point (best negotiated at low tide) the schists display at least two sets of minor folds, and Ordovician conglomerate rests unconformably upon the schists. The conglomerate is 30 cm thick and passes east into sandstone, about 70 cm thick, and thence into pelite. The sandstones and pelites are strongly bioturbated. The surface of unconformity strikes at 002° and dips at 80° west, and is therefore overturned.

Locality 9: Meillionydd

An old quarry [217288] opening onto the eastern side of the road 300 m south of the house of Meillionydd.

The gneiss exposed is variable in mineralogy and texture, for whereas biotite-rich areas are well foliated, quartzo-feldspathic areas are coarser-grained and only poorly foliated. The rocks are chopped about by many shear planes.

Permission is necessary to examine the exposures on the hill slope immediately east of the road at Meillionydd itself [217292].

Coarse-grained, plagioclase-hornblende gneiss with very coarse-grained feldspar-rich bands are exposed in the lowermost old opening. The banding of the gneiss strikes at 090–100°, and the dip is from 70° north to vertical. The gneisses are cut by later sub-horizontal felsic veins rich in coarse-grained, pink microcline. Irregularly shaped segregation patches of similar material occur in the gneisses.

A few metres further up the slope the gneiss becomes leucocratic. It is well foliated, hornblende-plagioclase gneiss, but which is now rather similar to the gneiss exposed in the old roadside quarry south of Meillionydd. The Meillionydd exposures are clearly migmatitic gneiss, and Shackleton has interpreted them as the migmatitic envelope to the Sarn adamellite.

EXCURSION 2. THE RHIW AREA

The purpose of this excursion is to examine the layered hornblende-picrite and hornblende-gabbro intrusion of Mynydd Penarfynydd, the related intrusions and the sequence of Lower Ordovician volcanic and sedimentary rocks and their deformation. The mineral assemblages developed in the basic rocks indicate that the rocks belong to the prehnite–pumpellyite facies of very low grade regional metamorphism. The succession is tabulated in figure 5 (column 1).

Transport between exposures is not essential.
Use the excursion geological map, figure 15.
References: Cattermole (1969), Hawkins (1970) and Matley (1932).
1:50 000 OS sheet No. 123; or 1:25 000 sheets Nos SH22 and 23.

Locality 1: Mynydd Penarfynydd

Parking is difficult, but a car can be left at the junction of the tracks at [224274]. Take the track to Penarfynydd farm and, having first obtained permission, continue along the track to Trwyn Talfarch. Descend over the boulder clay onto the beach. Proceed south over the boulders to the cliff of Ordovician shales.

About 30 m of sooty black, pyritous shales, mudstones and silty mudstones are seen beneath the overlying sill of layered mafic and ultramafic rocks. The beds contain thin, pale silty bands which are burrowed. Graptolites, including *D. bifidus* (Hall), indicate the *bifidus* zone of the Llanvirn. The beds dip at 30–35° towards 140°. About 10 m below the intrusion, the sediments become hardened and bleached by contact metamorphism and a metre below the junction they take on a crude columnar jointing and become spotted. The junction is difficult to reach unless it is low spring tide; at low neap tide some awkward scrambling is necessary. The contact is sharp and more or less concordant with the country rocks. The lowest 2 m (zone A of Hawkins 1970) is a very fine-grained, brown, chilled facies which has taken a poor cleavage. Above zone A, zone B is 11 m thick and consists of hornblende-dolerite passing up into hornblende-gabbro. Both zones A and B are thought to have undergone feeble, regional metamorphism which has reduced the plagioclase from labradorite or bytownite to albite-oligoclase.

Succeeding zones are best examined on the hill slope. Walk back to the drift and ascend the cliff to the feature on the hillside marking the base of the intrusion.

Rocks of the various zones recognised by Hawkins are well exposed and can be examined on an ascent of the hillside, but should not be

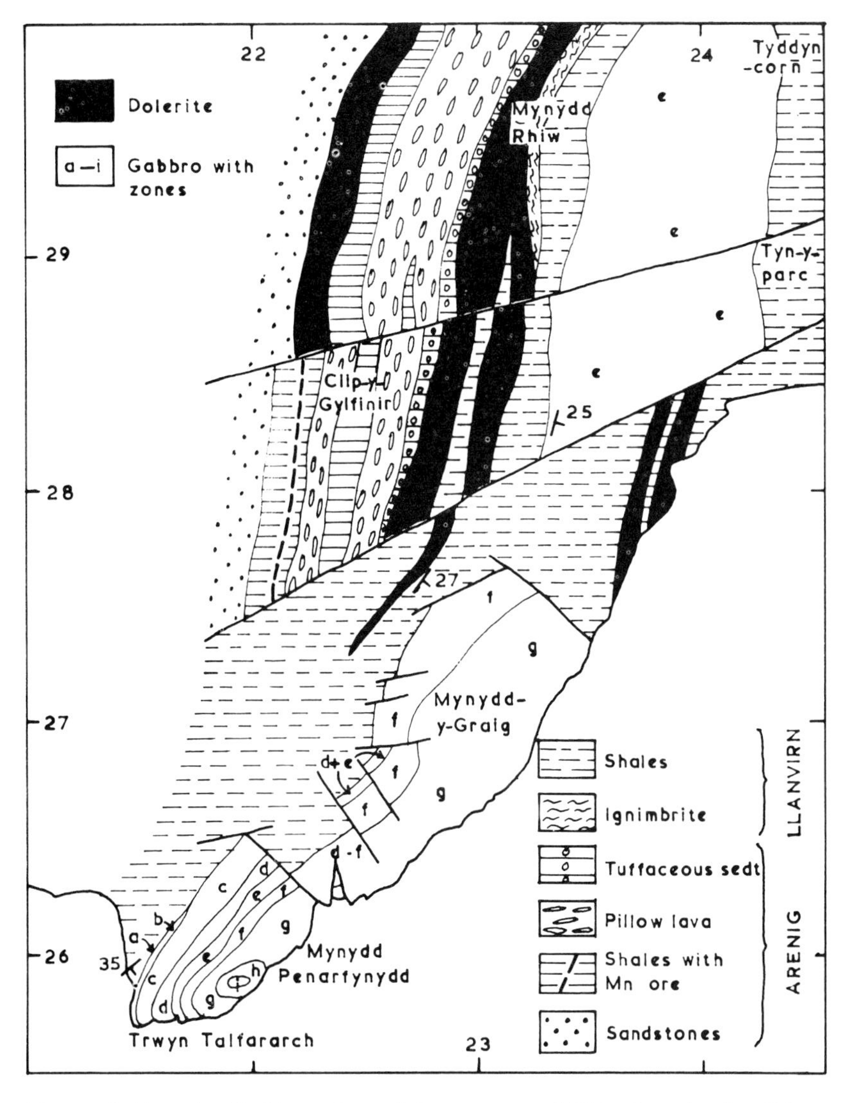

Figure 15. Excursion 2. The geology of the area around Rhiw. Largely after Crimes (1969a), Hawkins (1970) and Matley (1932), with permission.

hammered indiscriminately. Zone C consists of hornblende-picrites, which are cumulates and become rhythmically layered towards the top. This shows itself on weathered surfaces as ridges and grooves, and the rocks show honeycomb weathering. They are beautifully exposed on the high crags of Penarfynydd, and the bands are continuous laterally for up to 15 m. Features similar to sedimentary slump structures have been described by Hawkins. The succeeding leuco-gabbros of zone D are also banded, but they are very much paler than the picrites; they are more closely jointed, and do not weather into grooves and ridges. The upper layers of zone D have been disturbed and are truncated by the succeeding zone E, the pegmatitic gabbro, whose base is irregular. The rock is very coarse-grained (crystals of clinopyroxene are often 3 cm across). Zone F consists of very pale-weathering (off-white), highly altered gabbro. Zone G is a much darker, ore-rich, hornblende-gabbro. Zone H is a metadiorite and is thought to have formed near the roof of the intrusion because it has undergone mild regional metamorphism similar to the facies near the lower contact. Furthermore, the rocks are no longer cumulates. Finally, zone I consists of a coarse-grained metagranophyre mainly composed of albite and quartz, with some green amphibole.

Many of the plagioclase-bearing members of the intrusion show partial alteration of plagioclase to prehnite. Pumpellyite is uncommon, but the rocks now belong to the prehnite–pumpellyite facies of very low grade regional metamorphism.

The upper contact is nowhere exposed, but Llanvirn sediments can be examined on a steep slope south of Bytilith where they have been let down between a pair of faults.

Locality 2: Mynydd-y-Graig

Walk back northeastward towards Graig Fawr [224268].

A southeasterly traverse across Graig Fawr shows the succession to consist of zones D, E, F and G. No upper or lower contacts are exposed. The prominent crag of Cregiau Gwinen [228275–232276] consists of the pale-weathering gabbro of zone F. The Mynydd-y-Graig and Penarfynydd masses are thought to be part of the same intrusive sheet.

Locality 3: Clip-y-Gylfinir

Travel to [224281]. There is plenty of room to park a car at the verge, clear of the road. Cross the stile built into the wall and take the largely overgrown track leading to the disused manganese workings [222284] on the south-western slope of Clip-y-Gylfinir. Exposures occur in two small adjacent pits.

In the lower pit cleaved, brown-weathering mudstones with a thin chert can be seen. The bedding dips 60–70° towards 278°. A close fracture cleavage, S_1, dips 80° due east. Bedding and S_1 have been refolded by open, sideways-facing F_2 folds which plunge 20° due north. The folds are accompanied by a more or less horizontal, weak fracture cleavage, S_2, and its presence causes the mudstones to break up into pencil-like fragments. The manganese ore has been largely removed but, on the western side of the pit, chunks of the pisolitic and oolitic ore still adhere to the wall. Graptolites obtained from the mudstones on the eastern side of the pit indicate the presence of the *hirundo* zone. Small orthids can also be obtained here.

The smaller pit immediately to the north shows that the manganese ore is deformed, presumably in a disrupted F_1 anticlinal structure, since mapping indicates that the beds turn over to dip eastward beneath the succeeding rocks of Clip-y-Gylfinir.

The graptolitic and chert-bearing mudstones accumulated under quiet conditions presumably below wave base, but a brief period of shallowing and strong current action is indicated by the oolitic and pisolitic manganese ore.

Locality 4: [223284]

Some 40 m up the hill is a small crag. The exposure shows crudely pillowed, spilitic lava at the base overlain by massive, crudely columnar-jointed spilite. The pillows are amygdaloidal with chlorite, epidote and silica infillings, whereas the massive spilite is medium-grained and carries only chlorite amygdales. Blue-green chert occurs in some inter-pillow spaces. At the top of the hill, beneath the signal station, the rocks become pillowed once more. It may be that a single flow is exposed, the flow having a pillowed top and bottom, but a massive, columnar-jointed interior. Thin sections reveal prehnite and pumpellyite after plagioclase.

Locality 5: [226286]

Walk round the fenced-in signal station. More shales of the *hirundo* zone are concealed in the north–south-trending depression. Cross northeast to the low ridge with exposures of a second pillow lava. The pillows are strongly amygdaloidal with infillings of chlorite, calcite and silica. Thin sections show the same alteration of plagioclase to albite, prehnite and pumpellyite as in the first lava.

Locality 6: [229286]

Traverse due east for 350 m down the dip slope of the upper surface of the pillow lava and across an exposure gap. The gap conceals more shales, reputedly of the *bifidus* zone, but then a prominent crag of dolerite is reached. The rock is deeply weathered and columnar-jointed, and contains chlorite amygdales. The attitude of the columnar jointing indicates the sill is inclined to the east at about 30°.

Locality 7: [232293]

Follow the footpath northnorthwest towards Mynydd Rhiw. Cross the corner of the wall to reach the crags.

Just east of the northwest-trending wall is a grey-weathering, columnar-jointed dolerite with a thin upper surface of fine-grained, chilled marginal facies. The dip surface is overlain by a flinty, purple-welded tuff with strikingly obvious green fiamme. The rock is very fine-grained and may splinter dangerously when hammered. (In thin section this rock is seen to contain stilpnomelane.) Exposures are not good and may be missed unless care is taken. The presence of the ignimbrite indicates that subaerial conditions were established locally for a brief interval. The dip slope is succeeded by a small flat in which further shales of the *bifidus* zone are concealed. The shales are succeeded by the very coarse, hornblende-gabbro sill of Mynydd Rhiw. The rock type is the same as that of zone E of Penarfynydd and Mynydd-y-Graig. Thin sections of some specimens show extensive alteration of plagioclase to prehnite. Neither the top nor the bottom contacts of this gabbro sheet is exposed. The slope down to the drift-covered plain landward of Hell's Mouth is largely covered in gorse and huge boulders of gabbro. Traverse northeastward to the corner of the field and the road. Cross to the footpath running southeastward down the slope. Between Tyddyn-corn and Tyn-y-parc there are a number of small exposures of mildly contact-altered and bleached mudstones and slates of the *bifidus* zone, which must be close to the roof of the gabbro.

EXCURSION 3. THE CAMBRIAN AND ORDOVICIAN OF ST TUDWAL'S PENINSULA

The aim of this excursion is to examine the Cambrian succession and lithologies; the sub-Arenig unconformity; and the sedimentary facies of the Arenig. The deformation of the Arenig by the F_1 and F_3 phases is also examined. The succession is tabulated in figure 4 (column 4), and in figure 5 (column 2).

A single car can be left on the verge of the track 30 m west of the junction at [297242].
Use the excursion geological map, figure 16.
References: Bassett and Walton (1959), Crimes (1970a,b) and Nicholas (1915, 1916).
1 : 50 000 OS sheet No. 123; or 1 : 25 000 sheets Nos SH22 and 32.

Locality 1: Trwyn Carreg-y-tir [286240]

The small inlet has been eroded along a fault, visible in the cliffs and trending at 120°. Traverse upward from the seaward end of the headland across the Hell's Mouth Grits. The formation consists of interbedded greywacke sandstones, siltstones and mudstones. Greywacke sandstone units may be up to 4 m thick, but most are between 30 cm and 1 m and the thickness remains constant along the strike. Siltstones and mudstones tend to be from 15 to 50 cm thick. There is a wealth of sedimentary structures. Many sandstones are graded and completed by a parallel-laminated upper portion. Some consist of several graded units, whereas others show only the parallel-laminated interval. Sole markings include flute casts, groove casts and prod marks. Load casts and flame structures may be seen. Ripple-marked bedding planes are common. The greywackes are proximal turbidites. Crimes has reported greywacke-filled feeding burrows 0·5–1 cm in diameter, both in the mudstones and on the base of greywacke units. Directional structures indicate currents from a northerly direction. Composition of the clasts indicates a source area of metamorphic rocks like the Mona Complex of Anglesey.

Bassett and Walton obtained a Protolenid trilobite fauna indicating a Lower Cambrian age. The fauna has recently been described (Bassett *et al* 1976) and a late Lower Cambrian age confirmed. Other fossils include sponge spicules and inarticulate brachiopods.

The Mulfran Beds are exposed at the top of the cliff in a series of old workings. They consist of manganese-rich shales with mudstones, and greywackes usually about 50 cm thick. The greywackes are similar to those of the underlying Hell's Mouth Grits. The pelitic rocks are blue-grey on a fresh surface and many show ripple-drift bedding, but on weathering they develop a characteristic thick purple-black crust of manganese oxides.

Locality 2: Trwyn Cilan

Walk southeast along the cliff top over poorly exposed Mulfran Beds.

On the cliff top at [292235] exposures of Cilan Grits with current ripple-marked upper surfaces may be seen. Some of the grits are strongly conglomeratic and graded with quartz pebbles up to 1 cm across.

Immediately east of Trwyn Cilan [294231], it is possible to scramble down to the foreshore—care is needed. Here the Cilan Grits consist of bluish, conglomeratic greywackes, greywackes, siltstones and mudstones.

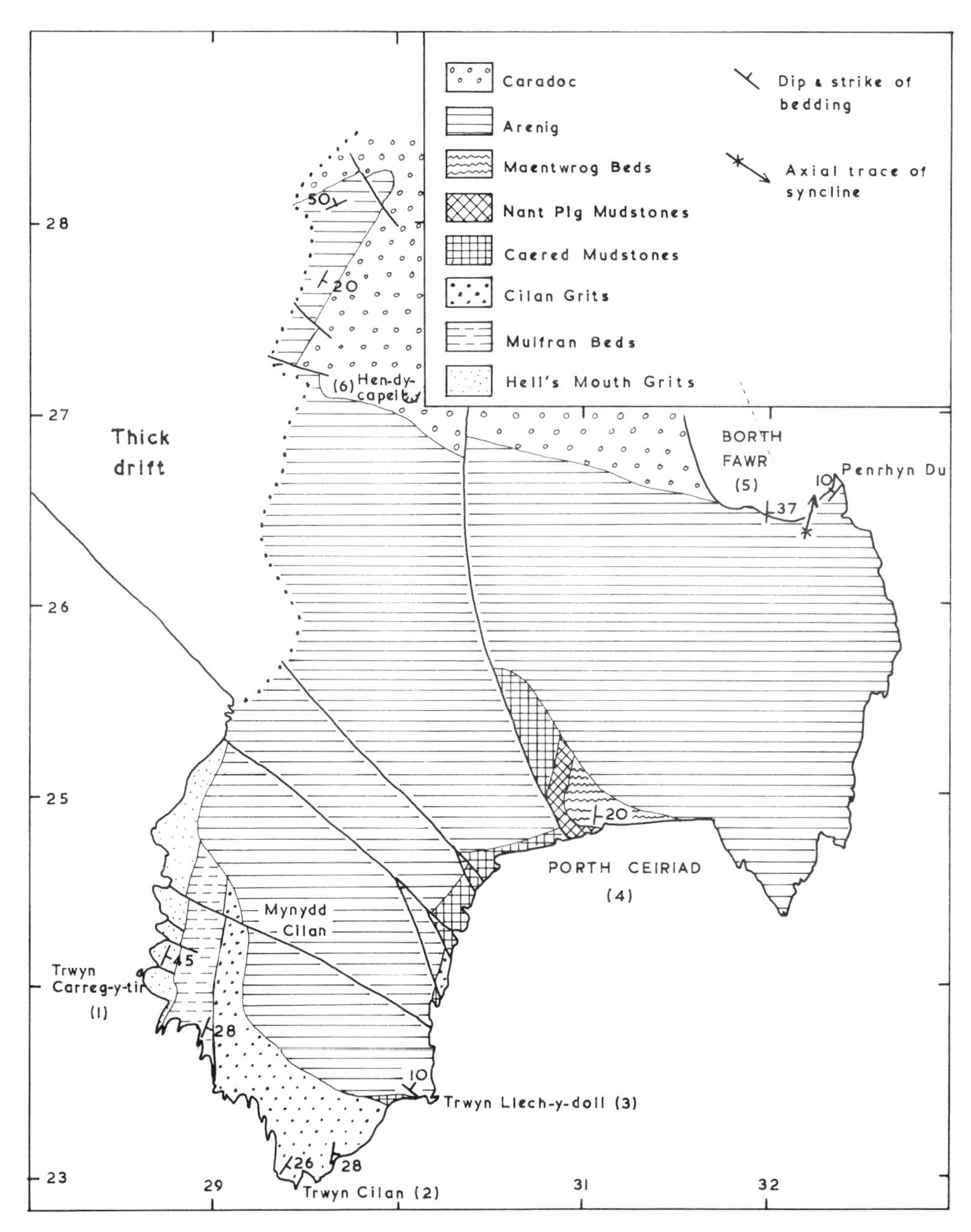

Figure 16. Excursion 3. The geology of St Tudwal's peninsula. Largely after Nicholas (1915) and Crimes (1969a), with permission.

Greywackes and conglomeratic greywackes commonly range from 50 cm–2 m in thickness. Sole markings include flute casts, groove casts and prod marks. Washouts are common. Most are graded and only a very few show the interval of parallel lamination. They are probably proximal turbidites. An interesting feature of several greywackes is the irregular upper surface, clearly erosional, upon which the succeeding unit of pelite rests. Presumably in these instances the erosive agent was a turbidity current which deposited its load further down the palaeoslope.

Slump features are common in the siltstone units. Towards the top, finer-grained greywackes and siltstones and blue, red and green, sandy mudstones are dominant.

Locality 3: Trwyn Llech-y-doll [302234]

Walk northeastward along the foreshore towards Trwyn Llech-y-doll and observe the sub-Arenig unconformity exposed in the cliff face and the recent landslip in the Arenig just beyond (figure 17).

On the way, notice the massive greywacke unit with the prominent quartz-filled tension gashes. Beneath the unconformity about 35 m of Lower Caered Mudstones are exposed resting upon an extensive pavement of topmost Cilan Grit. The topmost member of the Cilan Grits is a prominent, fine-grained sandstone with three sets of spectacular joints well seen in a pavement excavated along the upper bedding surface. The Caered Mudstones begin with laminated beds of bluish-grey-weathering mudstones which possess an obvious sub-conchoidal fracture. Interbedded grey mudstones, again with a sub-conchoidal fracture, contain clasts of biotite. An x-ray examination by R J Merriman of IGS (1976, personal communication) has indicated that the horizons probably represent former bentonites. Another horizon near the base is silicified and white-weathering, and is clearly a former vitric tuff. Siltstones and fine-grained sandstones showing ripple-drift bedding are present, and the upper surfaces of the beds are commonly ripple-marked. Sideritic concretions up to 30 cm across are developed in some of the siltstones and sandstones, and siderite is disseminated throughout most of the sediments. Beneath the surface of the unconformity, fine-grained, greenish white siltstones, with a conchoidal fracture, are again developed and probably represent further silicified vitric tuffs.

The basal Arenig conglomerate consists of two horizons, each rather more than 1 m thick, and separated by about 50 cm of interlaminated sandstone and mudstone. The lower conglomerate contains the nodular structure previously called *B. undosa* and pebbles of mudstone, jasper, quartz and quartzite. The interlaminated sandstone/mudstone horizon is strongly bioturbated. The basal beds are overlain by fine- to medium-grained sandstones interlaminated with mudstones. *Teichichnus* and *Phycodes circinatum* are common. Because the cliff is vertical, higher beds are best examined beyond the landslip, where well sorted sandstones contain abundant *Skolithos*. They are succeeded by a cross bedded unit 40 cm thick which is devoid of burrows; then come 1·5 m of interlaminated mudstones, siltstones and fine-grained sandstones, all of which are strongly bioturbated. The succeeding sandstones are cross bedded and are followed by interbedded sandstones (sometimes conglomeratic) and mudstones, all strongly bioturbated. Nicholas (1915) obtained *Dictyonema* from the mudstones. Higher units are exposed in the vertical, crumbly cliff and are inaccessible. Clearly the basal conglomeratic facies, with its evidence of strong current action, suggests accumulation under littoral conditions, whereas the higher beds, being finer-grained, well sorted and with abundant *Skolithos*, may indicate accumulation in a sub-littoral environment.

Locality 4: Porth Ceiriad [3024]

Return to the unconformity west of the Trwyn. Ascend to the cliff top by the path which runs along the unconformity. Care is needed, especially if the ground is wet. Walk north along the cliff-top path and observe the clean, cross bedded Arenig sandstones. Either walk around the cliff top to Porth Ceiriad [312248], or return to the transport and drive there. Cars may be left in the cliff-top car park. Descend to the beach.

The bay to the east is backed by cliffs cut in drift. Immediately to the west are exposures of the Maentwrog Beds, 33 m thick and dipping 18° towards 095°. The westward traverse therefore takes one down the succession. The highest beds consist of 10 m of black, blue-grey and green, laminated shales with thin siltstones and fine-grained greywackes 3–15 cm thick. They are underlain by about 10 m of fine-grained greywackes and siltstones 5–30 cm thick and with thinner shale partings.

Figure 17. The cliff immediately west of Trwyn Llech-y-doll, showing Arenig sandstones resting unconformably on Cambrian Caered Mudstones. In addition, note (i) the landslipped and rotated block of Arenig sandstone resting on a curved backscar, and (ii) the small, inclined normal faults in the Arenig sandstones forming the headland.

Finally, the basal bed is a 1 m thick, calcareous conglomerate. An abundance of sedimentary structures seen on wave-polished surfaces and bedding planes include convolute lamination, sandstone dykes, sand volcanoes, flute casts, groove casts, bounce and prod marks, load casts, ripple-drift bedding in siltstones, and a variety of ripple markings. Irregular burrows may be seen on the base of some of the siltstones. The lowermost greywackes are calcareous and sometimes concretionary. The basal, calcareous conglomerate contains pebbles of fossiliferous limestone, mudstone and rhyolite. This downward increase in the thickness and frequency of the greywackes indicates that turbidity current action progressively waned as the Maentwrog Beds accumulated and the turbidites became more distal in character.

The conglomerate is underlain by the Nant Pîg Mudstones. These are mainly dark blue and black, pyritous mudstones and siltstones, often finely laminated, and containing rare, thin, black limestones. Near the top, however, the proportion of siltstones is higher and occasional lenses of fine-grained sandstone about 3 m × 10–30 cm may represent channel fills. Some fine-grained sandstone bands have developed calcareous concretions. An abundant trilobite fauna indicates an age intermediate between the zones of *P. hicksi* and *davidis*. The beds are gently flexed and cut by a series of small normal faults (figure 18).

As the fault zones are approached, swarms of mesoscopic faults with throws of a centimetre or so affect the sediments. Folds plunge at 12° towards 015°, and an axial-planar cleavage is developed. The steep faults

Figure 18. The Nant Pîg Mudstones exposed in the cliff at Porth Caeriad. The parallel-laminated, black mudstones and siltstones are gently flexed and show development of an axial-planar cleavage. A vertical zone about 1 m wide has been weathered in to the left of the hammer.

trend between 285° and 330°; one low-angled fault trends at 010° and dips 30° to the east.

Continue westward along the foreshore across a pavement of Nant Pîg Mudstones, to a prominent fault trending at 340° and a large fall of blocks of Ordovician sandstones. Cross the cliff fall, which conceals Caered Mudstones and Flags. Observe the Ordovician sandstones in the upper cliff resting unconformably upon the partly obscured Caered Mudstones and Flags. The higher Caered Mudstones and Flags are purple, green, brown and grey, fine-grained sandstones, siltstones and mudstones with a sub-conchoidal fracture. Parallel lamination is usual. This group of sediments passes down into predominantly green siltstones and fine-grained greywackes which commonly are parallel-laminated, but occasionally show cross lamination and ripple-drift bedding. Bedding surfaces are sometimes ripple-marked. Several white-weathering, vitric tuffs are conspicuous. Burrows parallel to the bedding are common. An extensive Middle Cambrian fauna has been described by Nicholas (1915).

Locality 5: Porth Mawr [3226]

Drive to [322263] where a car can be left on the verge of this wide unmade road. Walk north down the road to the beach and along the foreshore to the old lifeboat slipway at Penrhyn Du.

Arenig beds are exposed along the shore in a major F_1 syncline. At the point, cleaved, bioturbated sandstones with thin mudstone partings form the eastern limb of the major syncline and dip 10° towards 324°. The cleavage, S_1, dips 60° towards 292°. *Fodinichnia* include *Phycodes circinatum* and *Teichichnus.* 30 m southsouthwest of the point a small fault trends at 355° and downthrows to the west. The downthrow beds consist of laminated and bioturbated, fine-grained sandstones and siltstones with thin mudstones. They are overlain by less obviously bioturbated, laminated siltstones and mudstones 1 m thick. 50 m from the point, cross a shallow fold-pair trending at 020° with zero plunge. Continue up the succession and about 65 m from the point, the sediments become finer-grained; the silty laminae become less prominent, and the proportion of mudstone increases. Bioturbation is less evident and the pelitic sediments are well cleaved (S_1) and are now slates. The sediments belong to the *Fodinichnia* Shaly Sandstone facies overlain by the Silty Mudstone facies of Crimes (1970b). Sub-littoral, quiet conditions probably prevailed. In a few metres there is another minor flexure of a monoclinal type. At 120 m from the point, both bedding and S_1 are re-folded by an F_3 structure. A close fracture cleavage, S_3, can be seen in the F_3 fold hinge. S_3 is vertical and trends at 270°. The fold brings in younger, bioturbated siltstones and fine-grained sandstones temporarily. 40 m northeast of the entry to the beach, the major F_1 synclinal axial trace is crossed and the slaty beds come in again. Beds now dip 30° towards 074°. A small, vertical fault trends at 292° and hereabouts S_3 is prominent as a vertical fracture cleavage striking at 250°. At the entrance to the beach, bedding becomes obscured, but both S_1 and S_3 are themselves deformed by late kink bands trending at 290°.

Cross westwards along the beach, over the drift-filled gap, to exposures of laminated and bioturbated siltstones. Bedding strikes at 010° and continues to dip east, now at 37°. The easternmost exposures are of interbedded, fine-grained sandstones, siltstones and thin mudstones, which are obviously bioturbated, and are the lowest beds exposed on the western limb of the F_1 major structure.

Locality 6: Hen-dy-Capel [300272]

In the two old pits south of the road, sediments low in the Caradoc are exposed. Two oolitic and pisolitic ironstones about 5 m thick are separated by about 1 m of black shales. The shales have yielded graptolites indicating the zone of *N. gracilis.* Nicholas (1915) thought these beds were thrust over the underlying Arenig, but Crimes (1969a) has suggested an unconformable relationship, citing the pisolitic ironstones as evidence of shallow water. The black ironstone consists of chlorite ooliths and pisoliths up to 2 cm across in a matrix of chlorite, siderite and magnetite. The sediments belong to the Oolitic Facies of Crimes (1970b).

EXCURSION 4. THE CLOSURE AND SOUTHEASTERN LIMB OF THE LLŶN SYNCLINE

In this excursion, the structure, succession and igneous rocks of the southeastern limb and closure of the Llŷn Syncline are examined. The sequence is tabulated in the Ordovician correlation chart in figure 5 (column 3).

Use the excursion geological map, figure 19.
References: Fitch (1967), Matley (1938) and Nicholas (1915).
1:50 000 OS sheet No. 123; or 1:25 000 sheets Nos 32 and 33.

Locality 1: Pen Benar, Abersoch [3128]

It is usually possible to park in or near Abersoch. Take the road that leads to Pen Benar and descend to the rocks before the new landing stage. Walk back westward to the first exposures (figure 20).

These are northwesterly dipping, grey mudstones with occasional beds of grey and brown, concretionary sandstones. The mudstones are underlain eastward by a white, highly feldspathic siltstone about 1 m thick, and then by grey mudstones and siltstones. Then come dark grey, finely laminated mudstones which weather reddish-brown, contain paler siltstone laminae and have yielded *Dictyonema*, and are of early Tremadoc age. A thrust dipping northwest at 35° brings the *Dictyonema* band over younger grey, green and brown shales, siltstones and fine-grained sandstones. The sediments are bioturbated and large sideritic concretions up to 1·5 m across occur in the coarser horizons. Minor F_1 folds verging to the southeast and plunging at 25–30° towards 025° may be seen in both the *Dictyonema* band and the bioturbated sediments, and an anticlinal hinge is crossed about 30 m before the new landing stage is reached. The beds now dip southeast at 50–60°. In addition to the minor F_1 folds, the

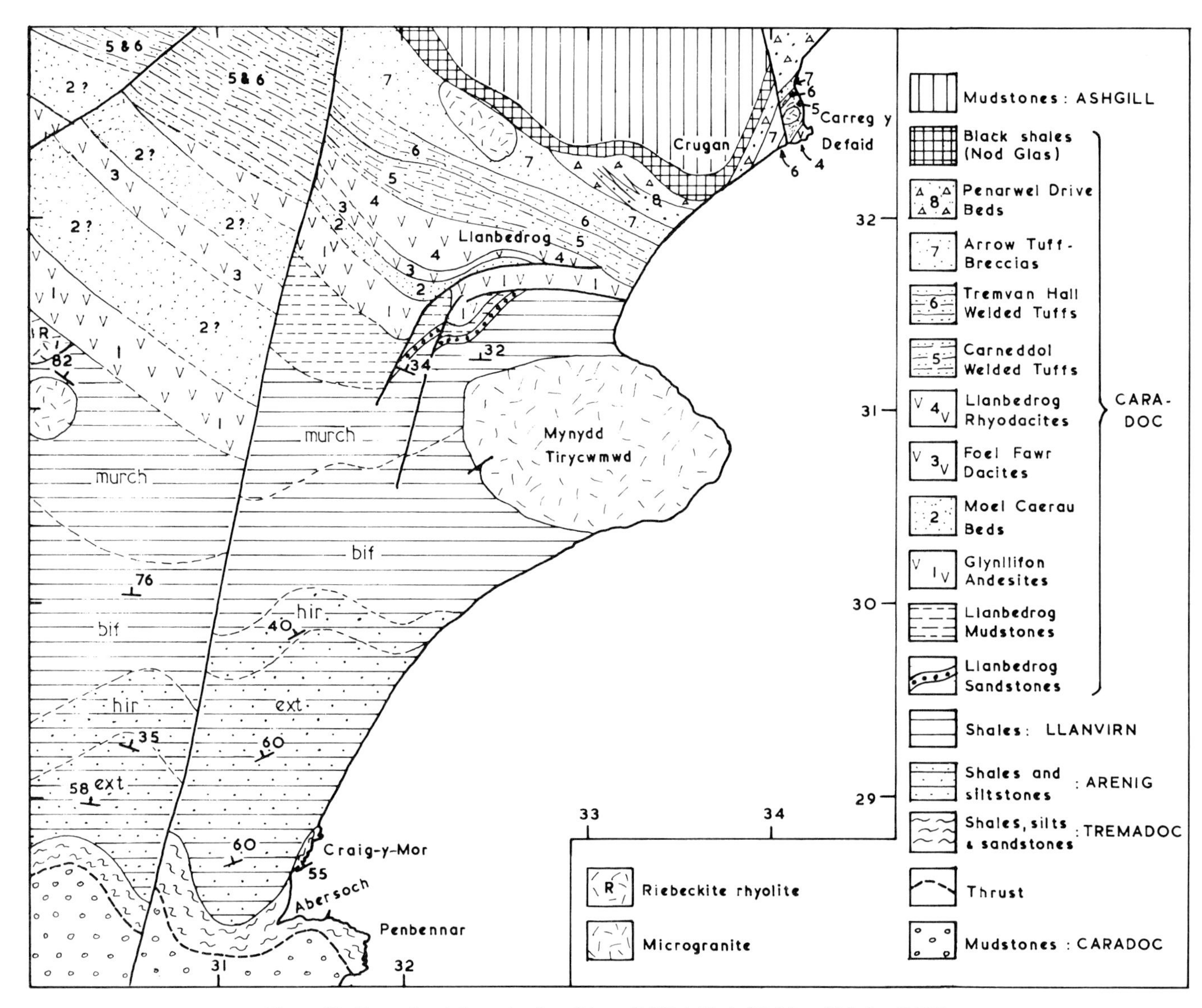

Figure 19. Excursion 4. Largely after Crimes (1969a), Fitch (1967) and Matley (1938).

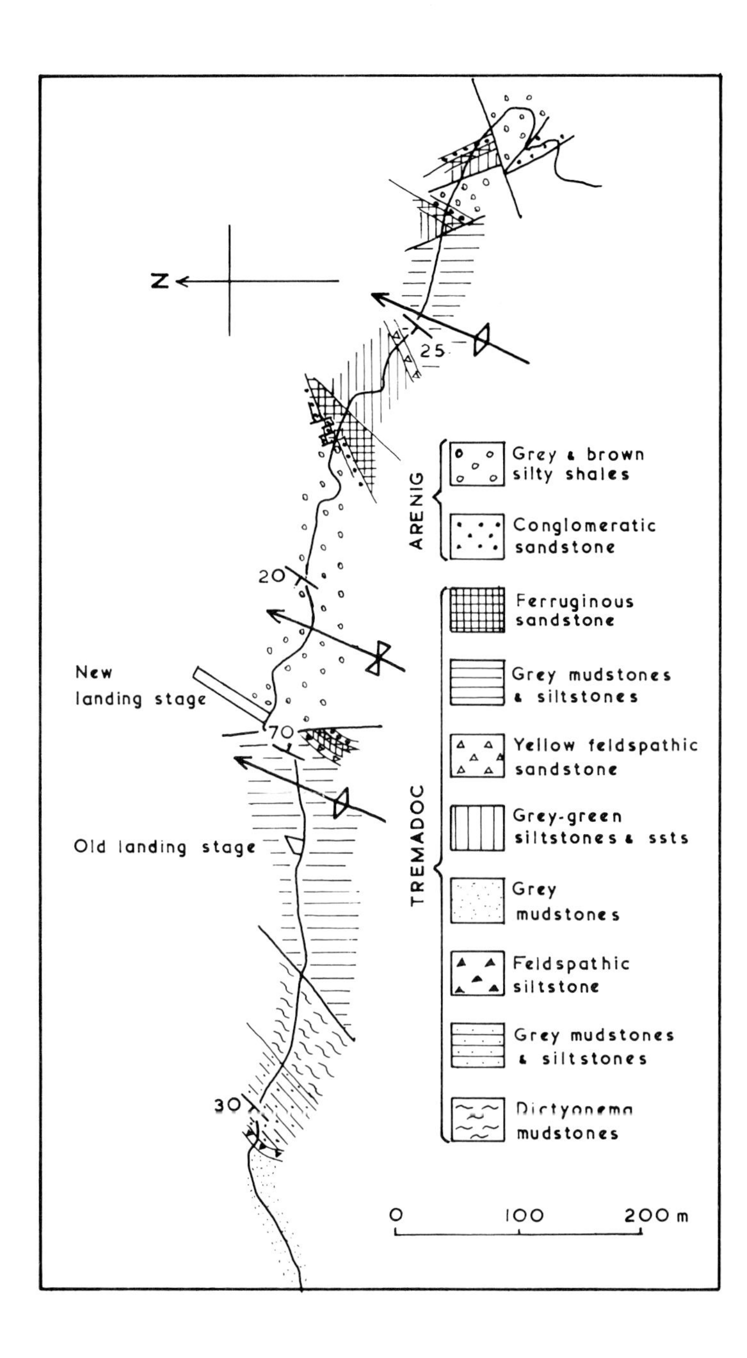

Figure 20. Detailed geological map of the Pen Benar foreshore, Abersoch. After Crimes (1969a), with permission.

more pelitic beds on the northwestern limb of the anticline also show a development of F_3 minor folds which plunge at angles of up to 35° towards 310°. An axial-planar fracture cleavage dipping at 80° southwestward is sometimes developed.

Immediately before the new landing stage, a vertical fault lets down the youngest beds seen in this traverse. They are bioturbated, grey and brown, laminated, silty shales and probably belong to the *extensus* zone of the Arenig. The beds occupy a syncline, the hinge of which is crossed some 70 m east of the landing stage. About 130 m east of the axial trace of the syncline, a 2 m thick conglomeratic sandstone containing mud pellets forms the base of the Arenig and dips northwest at about 30°. It is underlain by about 3 m of iron-rich, fine-grained, feldspathic sandstone ascribed to the topmost Tremadoc. Below come grey mudstones with bioturbated siltstone laminae, and then 4 m of yellow, feldspathic siltstones and shaly mudstones. Those are followed eastward by the same horizon of grey, green and brown, laminated and bioturbated, sandy shales and fine-grained sandstones seen earlier west of the new landing stage. Again, an anticlinal hinge is present in the beds, and the fold plunges at 25° towards 030°. Across the hinge the beds dip at angles of up to 45° towards the northeast and they are succeeded by grey mudstones with bioturbated siltstone laminae, then highly iron-rich, fine-grained sandstone marking the top of the Tremadoc, and finally the Arenig conglomerates at the eastern point.

According to Crimes (1969a), the Tremadoc and Ordovician rocks of Pen Benar, together with all the Ordovician rocks to the north, rest on a slide above the Caradocian rocks immediately to the south. The slide was invoked to explain the facies distribution and is thought to have been folded with the overlying and underlying beds. It is unfortunate that the slide itself is nowhere exposed.

Locality 2: Craig-y-Mor, Abersoch [315287]

Return to Abersoch and gain the beach north of the Afon Soch. Walk north along the beach to the prominent cliff exposures.

The beds dip northwestward at angles of up to 63°. Crimes (1969a) recognised high Tremadoc and basal Arenig sediments here. The grey, green and brown, laminated and bioturbated, silty mudstones and fine-grained sandstones form most of the exposures. They are overlain by the yellow, feldspathic siltstones and shaly mudstones, and in the south by grey mudstones with bioturbated, siltstone laminae. A thin basal Arenig conglomerate follows and contains the nodular structures formerly called *B. undosa*. The highest beds exposed are bioturbated, grey and brown shales.

The strata lie immediately east of the closure of the Llŷn Syncline and lie on the southeastern limb of the major structure. The folds of Locality 1 at Pen Benar represent minor folds on the same limb.

F_3 structures are very well developed: minor F_3 folds plunge at angles of 50° or so towards 315°, and are accompanied by an unevenly developed axial-planar fracture cleavage, S_3, dipping at 80° to the southwest.

Locality 3: Mynydd Tir-y-cwmwd

Return to the transport and drive towards Llanbedrog, but stop at about [322312].

The Arenig shales have been crossed and the Llanvirn shales are poorly exposed. They can be seen in poor exposures at [321308] to be finely laminated, soft, dark blue shales. They were probably laid down under quiet marine conditions below the reach of wave base.

Walk 200 m up the road to the track on the right leading to the old quarries cut into the south side of Mynydd Tir-y-cwmwd.

At [323307] the contact of the granophyre with the Llanvirn shales can be seen. The shales are mildly contact-altered, and against the contact have developed small albite porphyroblasts. Continue southeastwards to the old quarry at Tŷ'n-towyn. The granophyre shows several sets of cooling joints and a set more or less parallel to the topographic surface which are presumably due to unloading. The rock is a feldspar-porphyritic microgranite which, in thin section, shows a well developed granophyric texture.

Return to the main road. At [322314] on the western side of the main road the Caradoc succession begins abruptly with about 20 m of massively bedded, feldspathic sandstones exposed in an old quarry which, unfortunately, has recently been fenced off. The beds dip at about 35° towards 015° and are close to the hinge of the Llŷn Syncline. The massive sandstones are coarse-grained and up to 3 m thick near the base, but become finer and calcareous upward. Discontinuous conglomeratic horizons contain clasts of pelite, quartzite and gneiss. They are also exposed at [320313] on the valley side and the exposure, which is fossiliferous, can be reached from the footpath which runs southwest from Bodwrog. The sandstones are separated by thin mudstones and become thinner-bedded and finer-grained upward. They are overlain by the 80 m thick Llanbedrog Mudstones which yield an abundant brachiopod–trilobite fauna, as well as corals and molluscs. These brown-weathering, blue, calcareous mudstones and siltstones are exposed along the roadside for about 175 m. **Take care not to leave debris if fossils are collected.** The massive, conglomeratic, feldspathic sands imply a change from below wave base to shallow-water marine conditions. The sands contain a Soudleyan stage Caradocian fauna, and the fact that the Llandeilo stage has not been proved indicates a disconformity at the base of the sands.

A narrow lane leads westward from the bend in the main road beyond which is a development of columnar-jointed andesite lavas: the Glynllifon andesites. These are about 150 m thick and are succeeded by the poorly exposed Moel Caerau Beds which are said (Fitch 1967) to consist of about 80 m of bedded tuffs, tuffaceous sandstones and mudstones. The overlying Foel Fawr dacites can be examined alongside the footpath which leads north from the roadside opposite the boatbuilder's workshop. It is a vaguely flow-banded, feldspar-porphyritic rock, blue-green to blue-grey on a fresh surface and in the hand specimen is indistinguishable from the Glynllifon andesites. Return to the main road and walk east to the junction with the Mynytho road. Roadside cuttings are in the Llanbedrog rhyodacites. This is a very strongly feldspar-porphyritic rock, but it too has the familiar grey-green base.

Locality 4: Llanbedrog

Walk up the Mynytho road as far as the junction with the footpath on the right which runs alongside the school driveway [326319].

A recent roadside excavation has revealed a strongly eutaxitic, rhyolitic, welded tuff—the Carneddol Welded Tuffs. Take the footpath north across the Carneddol Welded Tuffs which form the ridge on which the school is built, and then through the next ridge to the north formed by the overlying Tremvan Hall Welded Tuffs. The footpath then skirts the southern boundary of the Castell-grug intrusive rhyolite which is emplaced in the Arrow Tuff-breccias. These are about 200 m thick and consist of blocks of rhyolite and pre-existing tuffs in a crystal-vitric tuff matrix.

The largely volcanic succession seen at Localities 3 and 4 succeeds shallow-water sediments. The rapid accumulation of volcanic rocks led to emergence and the establishment of short-lived subaerial conditions under which the welded tuffs accumulated.

Locality 5: Crugan

Return to the transport and drive to [335323]. Walk northwest along the track past Crugan for a little over 200 m.

Exposures of grey mudstone are seen adjacent to the track. They are grey, unlaminated and concretionary and yield a trilobite–brachiopod fauna. The fossils are very small and indicate the presence of the zone of *Phillipsinella parabola* of the Ashgillian.

Locality 6: Carreg y Defaid [3432]

Return to the road and proceed to the end of the cul-de-sac [341327] near Carreg y Defaid.

In a small quarry next to the lane blue-black, silty shales on the southeastern limb of the Llŷn Syncline strike at 030° and stand more or less vertically. The beds, correlated with the Nod Glas sediments, occur beneath the Crugan Mudstones of exposure (5) and graptolites indicate an horizon near the top of the zone of *D. clingani*. The beds overlie the volcanic succession and indicate an abrupt return to quiet conditions of accumulation below wave base. The Ashgillian mudstones of Crugan which succeed the *clingani* shales also accumulated under quiet marine conditions, but the two formations are separated by a small hiatus.

Walk to the beach and then southward to the first exposures. They consist of 6 m of bedded lapilli tuffs of the Arrow Tuff-breccia formation dipping 80° towards 320° and so clearly underlie the *clingani* shales of the previous exposure. They are underlain by bedded tuffs and agglomerates with blocks of rhyolite up to 2 m across, but the bedding becomes obscure and the agglomerate is cut out in places by veins of green and grey, rhyolitic, lapilli-rich, crystal-vitric tuff. Towards the base, the blocks themselves consist of welded, eutaxitic tuff set in a welded-tuff matrix.

A series of welded eutaxitic tuffs with interbedded non-welded lapilli tuffs underlies the agglomeratic, welded tuff. The tuffs have been ascribed to the Tremvan Hall Welded Tuffs and to the Carneddol Welded Tuffs. A massive, unbanded rock with sporadic nodules up to cannon ball size occurs next. The rock contains feldspars up to 10 mm across and has been interpreted as the nodular facies of a rhyolite plug by Fitch (1967). It is followed to the south by a faintly banded, orange-weathering rhyolite which can be seen in the low cliff to form a dome. A nodular marginal facies is very well developed about 10 m to the southeast of the centre of the dome, and dips at 45° to the southeast. The nodular facies is ovelain to the southeast by a flow-banded, flowage-folded and autobrecciated outermost facies. The point of Carreg y Defaid consists of a massive, agglomeratic, brecciated rock, ascribed by Fitch to the Llanbedrog Rhyodacites. Around the point a vertical dyke of tuff-breccia can be seen trending at 050°. The final exposures are of a beautifully eutaxitic, orange-weathering, welded tuff of the Tremvan Hall Welded Tuffs, dipping at 70° towards 290°. The degree of welding decreases upward so that the highest exposure (against boulder clay) is only slightly welded.

EXCURSION 5. THE CLOSURE AND NORTHWESTERN LIMB OF THE LLŶN SYNCLINE

The purpose of this excursion is to examine the sediments, volcanics, intrusive rocks and the tectonics of the overturned northwestern limb. The sequence is largely that tabulated in the Ordovician correlation chart figure 5 (column 3).

Use the excursion geological map, figure 21.
References: Matley (1938), Matley and Heard (1930) and Tremlett (1962).
1:50 000 OS sheet No. 123; or 1:25 000 sheets Nos SH23 and 33.

Locality 1: Mynytho Common

Drive to Foel Gron. There are no parking problems.

Examine the Foel Gron aplite exposed in the small quarries cut into the southern margin of the intrusion at about [302309]. The intrusion is probably a cylindrical plug. In hand specimen the rock is a pink microgranite with an aplitic texture. Walk north across the hill to the lane west of the school. Here the blue-black *bifidus* shales into which the aplite is emplaced are exposed and virtual contacts can be examined. There is little or no contact metamorphism.

Follow the track leading westnorthwest for about 400 m until outcrops of flow-banded and flowage-folded, riebeckite-bearing rhyolite are met with. Further northwest the rock is slightly coarser-grained and vesicular; dark blue riebeckite can be seen with a hand lens. The rock probably forms a sill which, mapping suggests, strikes at about 300° and dips 30° to the northeast.

Walk eastnortheast across drift-covered ground to a prominent crag at [302316]. The crag is made of flow-banded andesite lavas which, on fresh surfaces, are seen to consist of orange feldspar phenocrysts in a microcrystalline green-grey base. They probably represent the Glynllifon andesites of Llanbedrog and form a prominent feature trending northwest.

Locality 2: Foel Fawr [306321]

Return to the transport and drive to a point just south of the bend in the road at [304322] where a car can be left on the roadside verge. Climb the hill to the restored windmill on the summit.

Examine the exposures of the Foel Fawr dacites: they are flow-banded and autobrecciated. On a fresh surface they are seen to consist of phenocrysts of feldspar in a microcrystalline green-grey base.

Walk north for 200 m from the windmill across the dacites and an exposure gap to the next exposures which are of an agglomeratic, andesitic tuff containing blocks of rhyolite up to 15 cm across and lapilli of shale, andesite, rhyolite and feldspar crystals set in a vitric matrix. The tuff appears to fine upward and becomes stained purple. It is overlain by a purple, feldspar-porphyritic andesite.

Locality 3: Carneddol [302332]

Either walk across Carneddol by way of the footpath and farm tracks, or return to the transport and drive to [305327] where a car can be left at the roadside verge. Walk up the southeastern flank of Carneddol by way of the gates in the fields.

The hill consists of a pink-weathering, columnar-jointed, rhyolitic, welded tuff with a prominent eutaxitic structure. The eutaxitic structure dips 35° towards 045°. These are the Carneddol Welded Tuffs of Fitch (1967) and are at least 200 m thick. There are probably several flows present. The welded tuffs indicate that subaerial conditions had been established.

Locality 4: Inkerman Bridge [288328]

Return to your transport and drive to Inkerman Bridge.

A sill of porphyritic albite-granophyre is worked in the quarry on the western side of the valley. It is also exposed in the old quarry on the eastern bank. The lower marginal facies is a spherulitic felsite. The sill is of the order of 200 m thick and is emplaced into Llanvirn *bifidus* zone blue-black shales.

Locality 5: Garn Fadrun to Moel Caerau

Drive to Garn [278346]. Parking is difficult. Walk northward up to the end of the track which starts immediately east of the church. The footpath off to the eastnortheast is overgrown and impassable, so take the new track which runs northeast from the end of the lane and ascend part of the way up Garn Fadrun.

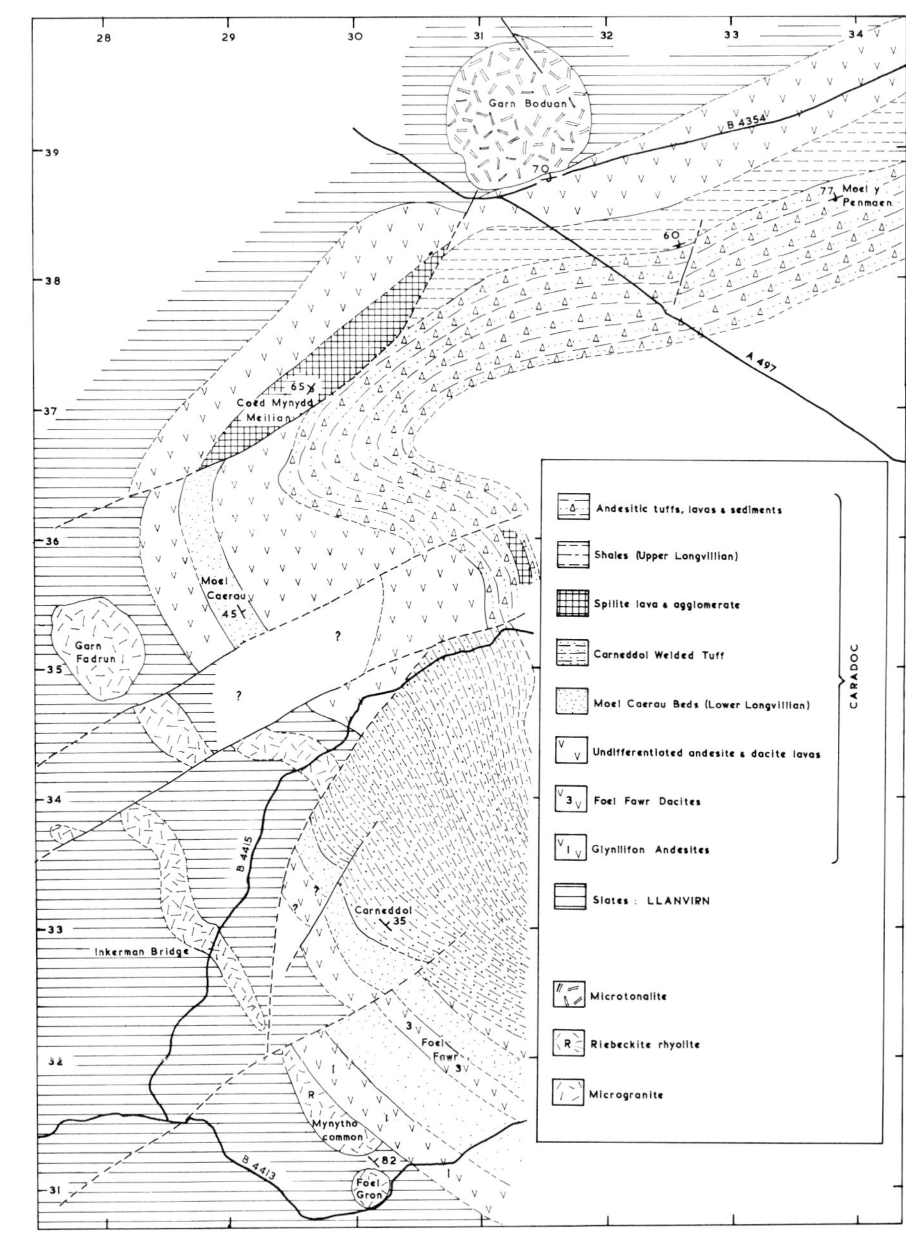

Figure 21. Excursion 5. Largely after Crimes (1969a), Matley (1938) and Matley and Heard (1930), with permission.

Mapping suggests the intrusion is cylindrical, having the form of a small boss. The rock is fine-grained, feldspar-porphyritic, pyroxene-granophyre, and is greenish grey on a fresh surface.

Walk east to the valley separating Carn Bach from Carreg Dinas and at [287350] blue-black shales in a small quarry have yielded to Crimes (1969a) graptolites which may indicate the zone of *G. teretiusculus*. The shales are indistinguishable lithologically from shales of proven Llanvirn age. Walk north onto Carreg Dinas. Examine the exposures of strongly vesicular andesite lavas. Continue northeast towards the summit of Moel Caerau. At [291356] sandstones and finer-grained sediments dip 45° towards 060°. The beds yield a shelly fauna of Lower Longvillian age. These are the Moel Caerau Beds of Fitch (1967) and must be about 75 m thick. The cross bedded sands indicate current action and the sediments are probably sub-littoral. The summit and the slopes to the north and east of Moel Caerau show exposures of feldspar-porphyritic andesite and dacite, which are often flow-banded, sometimes autobrecciated, and vesicular. They may be of the order of 500 m in thickness.

Locality 6: Coed Mynydd Meilian

Return to the transport and drive to an old roadside quarry at [297373].

The quarry is largely overgrown and is cut in spilite. Enter the larch wood near its southern boundary and exposures of spilite are seen on the right. They are flow-banded and amygdaloidal, and the flow planes dip 65° towards 308°. The beds are almost certainly overturned. In the south, where the wood protrudes into the arable land, is a low cliff with a superb exposure of crudely bedded, brecciated spilite with blocks up to 75 cm across in a matrix of strongly vesicular spilite. The beds are vertical or dip steeply northwest and probably lie stratigraphically above the spilite lava immediately to the north. The outcrops now lie in the steep to overturned northwestern limb of the Llŷn Syncline.

Locality 7: Garn Boduan

Drive to the old quarry at [315387].

On the northeastern side of the quarry are a series of bedded andesitic tuffs, often lapilli-rich, interbedded with green crystal-vitric tuffs which, in the hand specimen, look very much like feldspar-porphyritic andesite lavas. Bedding dips 70° towards 348° and is probably inverted so that the outcrops are well into the overturned northwestern limb of the major syncline. The tuffs dip steeply towards a sub-vertical wall of coarse breccia which can be examined here, but only after a scramble. It is better reached by ascending for a short distance the path which leads diagonally up the hillside from the western end of the quarry and then striking up through the brambles. The vertical wall of breccia can be seen to consist of a marginal brecciated facies of a feldspar-porphyritic, fine-grained rock. The rock rapidly becomes homogeneous inward. It weathers buff to grey and on a fresh surface is seen to consist of a greenish grey microcrystalline matrix supporting greenish black chlorite after pyroxene or amphibole and phenocrysts of brown plagioclase. It is a microtonalite and is regarded by the author as a plug-like intrusion with essentially vertical contacts. This interpretation is different from that of Matley and Heard (1930) who regarded it as a thrust sheet of keratophyre (= andesite).

Locality 8: Moel y Penmaen

Drive to [339389] and park on the verge. A track leads south to an old quarry cut in the northeastern angle of the hillside.

The exposed rock is a blue-grey, compact, feldspar-porphyritic, and sparsely amygdaloidal andesite.

Walk west to about the mid-point along the foot of the north-facing slope of the hill and climb up over the andesites. Rubbly tops and bottoms indicate a steep northerly dip to the andesite flows. Carry on southward and, where the exposures are restricted by fields of pasture on the west, the strongly amygdaloidal lavas (with chlorite amygdales) give way to well bedded, parallel-laminated, andesitic tuffs. Bedding dips 77° towards 342°, whereas the cleavage (S_1) dips less steeply at 73° towards 352°. The beds are therefore overturned, as indeed so too are the beds of exposures 6 and 7. Localities 6, 7 and 8 are therefore in the inverted northwestern limb of the Llŷn Syncline. A small quarry cut in the tuffs and tuffaceous sandstones and siltstones shows bands rich in fragmental trilobites and brachiopods. Other horizons are sometimes pebbly and conglomeratic, and there is plenty of evidence of shallow-water conditions of accumulation.

The fossiliferous horizons are stratigraphically overlain to the south (remember the beds are inverted) by a bedded rhyolitic agglomerate with rounded clasts of rhyolite and andesite in a rhyolitic tuff matrix. The clasts range from 10 cm across, down to the smallest fragments of the matrix. Bedding dips 74° towards 334°. The last well exposed rock in this ascending sequence is of a flinty, white-weathering, eutaxitic, rhyolitic welded tuff. The rock splinters when hammered and is blue-grey on a fresh surface. Clearly, therefore, at this locality the exposed sequence ends with the establishment of subaerial conditions, although no doubt these conditions were short-lived.

EXCURSION 6. TREFOR TO NEFYN

The aim of this excursion is to examine the succession and structure on the northwestern limb of the Llŷn Syncline and to examine the small plutonic intrusions. The stratigraphical succession is probably as shown in the table below.

Caradoc	Afon Erch Tuffs
Llandeilo	? Absent
Llanvirn	Blue-black slates Bioturbated khaki and olive mudstones and siltstones Blue-black slates
Arenig	Slates with silty laminae Pisolitic Ironstone Trwyn-y-tâl Sandstones

Use the excursion geological map, figure 22.
Reference: Tremlett (1962).
1:50 000 OS sheet No. 123; or 1:25 000 sheet No. SH34.

Locality 1: Trwyn-y-tâl

Drive to Trefor and park in the car park at [375473]. Walk west to the exposures at the eastern end of Trwyn-y-tâl [372475].

At the eastern end of the shore exposures the rocks are fine-grained, laminated, grey and buff sandstones, siltstones and silty mudstones, all of which are strongly bioturbated. Near the prominent adit the beds dip 85° towards 205°. Cleavage (S_1) is vertical and the beds presumably therefore face south and are the right way up. This is confirmed by the observation that laminated sandstones fill pockets scoured in bioturbated sandstones on the northern side of the adit. Cleavage/bedding lineations plunge at 33° towards 277°. The adit has been cut in an 8 m thick ironstone consisting of pisoliths and ooliths of a chlorite set in a black, fine-grained matrix of chlorite, siderite and magnetite. It is very pyritous. The ironstone is overlain on the south by blue-black mudstone with silty laminae which at [368472] strike at 292° and dip 58° southwest. The sediments are almost certainly Arenig in age.

30 m along the foreshore the bedding in bioturbated sandstones and siltstones strikes consistently at 044° and dips 55° northwest, so that an anticlinal axial trace has been crossed. However, where exposures are first met with on the shore, both bedding and S_1 can be seen to be re-folded about mesoscopic F_3 folds and a strong, vertical fracture cleavage, S_3, strikes at 100°.

Walk along the cliff top to the headland at [365474]. Bedding in fine-grained sandstones, siltstones and silty slates dips 50° towards 332°; a slaty cleavage, S_1, dips 55° towards 340°. The beds are the right way up and are in the northwestern limb of the F_1 anticline. The axial trace runs westsouthwest immediately north of the adit and is marked on the map (figure 22).

Carry on to the end of the cliff and descend to the shore just before the junction of the solid rock with the drift. As you descend, both bedding and S_1 can be seen to be folded into a series of F_3 folds plunging at 55° towards 277°. A well developed axial-planar fracture cleavage, S_3, is present.

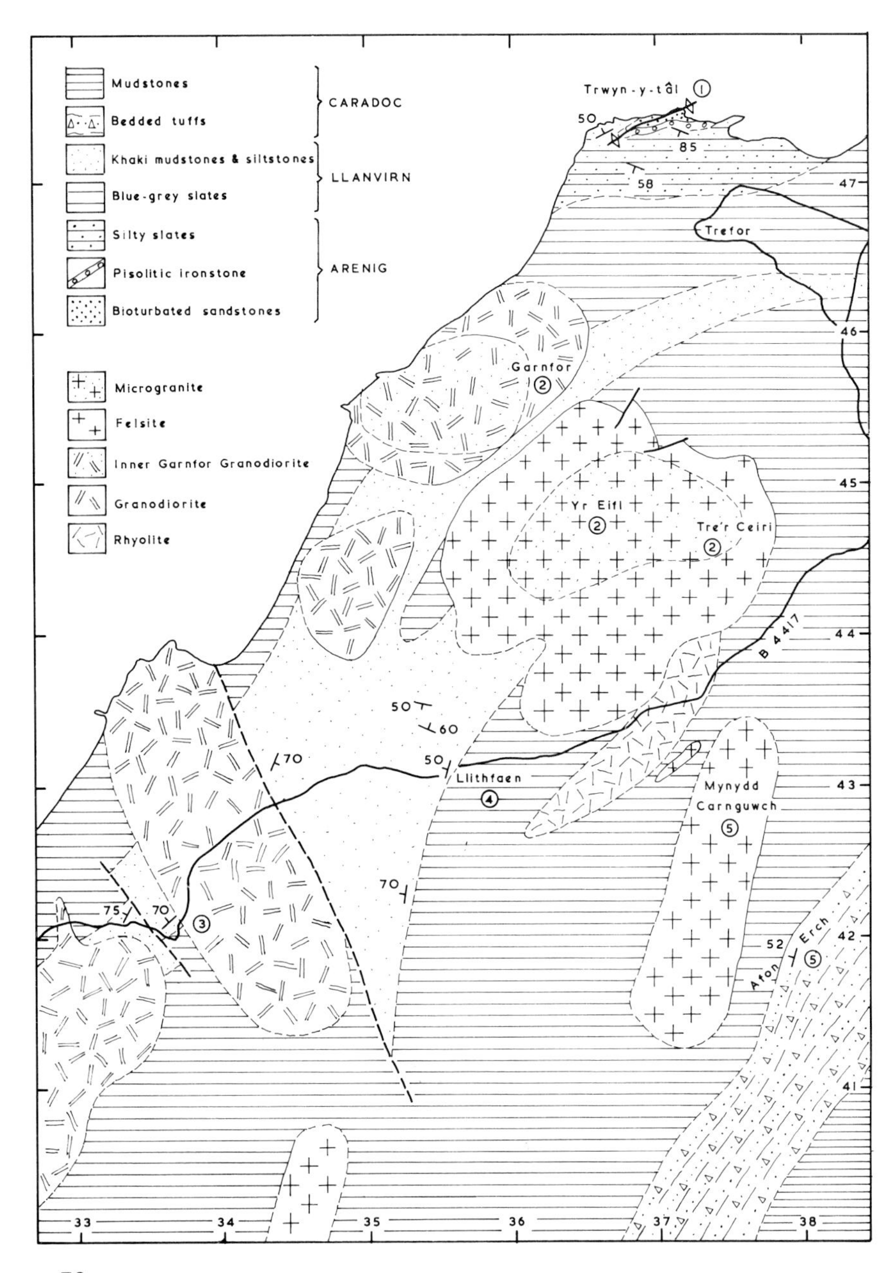

Figure 22. Excursion 6. Modified from Tremlett (1962), with permission.

The sandstones of Trwyn-y-tâl at the base of the succession probably represent the *Fodinichnia* Sandstone facies of Crimes (1970b) and were laid down under fairly quite marine conditions. The overlying oolitic and pisolitic ore indicates a temporary shallowing, but the succeeding silty slates belong to the Silty Mudstone facies and indicate a return to deeper water and quiet conditions.

Locality 2: Yr Eifl and Garnfor

Return to Trefor and drive through Llanaelhaearn and Llithfaen to Bwlch Yr Eifl [362454]. There is ample parking space on the southeastern side of the track.

Walk northeastward down the track to [363455] where bioturbated, green and olive silicified mudstones and siltstones are exposed at the trackside. These sediments have been mapped as rhyolites by Tremlett.

The Garnfor microgranodiorites can be examined in old quarry workings immediately to the northwest. The outer facies is worked here and it is a white or pinkish rock with fresh feldspars up to 3 mm across and smaller but conspicuous hornblende, biotite and magnetite in a fine-grained matrix. Xenoliths of a basic facies are common.

Walk southwestward back up the bwlch to small crags just to the northwest of the summit of the bwlch. The exposures are banded, blocky, khaki-weathering, silicified mudstones. The occasional burrows indicate their sedimentary nature.

Follow the track southwest to the fork at [359452]. Leave the track and follow the stream in the deeply incised valley for 100 m. The sediments are well exposed near the footbridge which crosses the stream. Once again their bioturbated character indicates they are sediments rather than rhyolites.

Follow the footpath nothwestward for about 500 m to one of the old quarry workings in the inner facies of the microgranodiorite. The facies is grey and microcrystalline, and ferromagnesian minerals are inconspicuous.

Return to the bwlch and climb Yr Eifl. The microgranite forming the summit of the hill is a grey to buff rock consisting of small feldspar phenocrysts in an aplitic matrix. The outer felsitic facies is well exposed in the northeasterly facing crags where, at [372451] the microgranite can be seen to be chilled against the felsite.

If time allows, it is very well worth while crossing the col to Tre'r Ceiri, where a magnificently preserved fortified village is sited on the same microgranite as Yr Eifl.

Tremlett regarded the Yr Eifl group of boss-like intrusions as Caledonian in age. No evidence was offered and the suspicion must be entertained that they are Ordovician in age.

Locality 3: [337420]

Drive to [337420] where sediments are exposed in a road cutting on the northern side of a tight bend in the road.

The sediments are greyish-black, slightly khaki-weathering, silty slates. The exposure is structurally very interesting. Over much of the exposure bedding dips 70° towards 315°, but is deformed by a series of sideways-facing F_2 crenulations, the axial surfaces of which dip 50–30° towards 315°. Both bedding and the F_2 crenulations are further deformed by a well developed set of small-scale F_3 folds which plunge at angles of up to about 70° towards 330° to 335°.

The beds are structurally, and presumably also stratigraphically, overlain by olive coloured siltstones. These are exposed in the roadside and the fields to the north.

Further north, to both west and east of the road, granodiorite is exposed. It is a pinkish buff, feldspar-porphyritic rock of medium to coarse grain size with abundant, dark greenish black hornblende.

Locality 4: Llithfaen

Drive northeast to Llithfaen.

On the north side of the main road, 50 m west of the crossroads in the village centre, are exposures of khaki, laminated siltstones. Bedding dips at 50° towards 280°. A cleavage, which presumably is S_1, dips 80° towards 320°.

To the northnorthwest, on the hill, is an extensive area of outcrops which have not been mapped by Tremlett. The rocks (brown- and khaki-weathering siltstones) can be reached by taking the road northwest from the centre of Llithfaen for 750 m, and then taking the footpath onto the hill from the cottages. They are folded into a series of southwesterly plunging F_1 folds with axial surfaces dipping steeply northwest. An anticlinal hinge is well exposed in the low crag just above the wall north of the house at [353433] and plunges at 20° towards 230°. No fossils have been obtained from the rocks as yet.

Locality 5: Mynydd Carnguwch and the Afon Erch

Drive to [372436] and park in the lay-by on the south side of the road.

Continue on foot towards the top of the pass and examine exposures of rhyolite in the rough pasture south of the road. This unprepossessing, sparsely feldspar-porphyritic, felsitic rock was interpreted by Tremlett as an Ordovician rhyolite.

Walk southeast to the summit of Mynydd Carnguwch [375429] and examine exposures of the porphyritic felsite. It is indistinguishable from the felsite of the flanks of Tre'r Ceiri, but the jointing suggests it may be part of a sill-like sheet inclined south or southeast.

Descend south to the track and follow the paths to the church at [374418]. Below the church, the Afon Erch has cut into the felsite sheet. Walk east along the north bank to [376417], where soft blue-black slates are exposed in the south bank of the stream. Bedding and cleavage appear to be coincident, dipping 80° towards 292°. No fossils have been obtained but the slates are probably Llanvirn in age.

Continue northeast to [377417] where an orange-weathering, rhyolitic welded tuff is exposed on the south bank. It has a fine eutaxitic fabric and consists of feldspar phenocrysts in a very fine-grained matrix.

At [379419] in the stream bed and in the south bank, deeply weathered, bedded, crystal-lithic tuffs are exposed. They are strongly cleaved and folded, and contain large feldspar crystals and lithic fragments of lapilli size. Cleavage dips 65° towards 302°; the folds plunge at 25° towards 235°. The tuffs are Caradocian, although no fossils have been obtained from this locality.

The purpose of this excursion is to examine the Caradoc, Ashgillian and Llandovery successions and to study the structure on the southeastern limb of the Llŷn Syncline. The sequence is tabulated in the Ordovician correlation chart in figure 5 (columns 3 and 4).

Use the excursion geological map, figure 23.
References: Harper (1956), Matley (1938), Roberts (1967) and Tremlett (1965).
1:50 000 OS sheet No. 124; or 1:25 000 sheets Nos 33 and 43.

Locality 1: Pen-yr-allt

Parking is no problem in Pwllheli. Take the Llannor road northwards from Pwllheli centre. As you ascend the hill out of Pwllheli there are more or less continuous exposures on the eastern side of the road.

The lowest exposures are of about 3 m of blue-grey, brown-weathering blocky mudstone. Bedding dips at 60° towards 340° so the exposures are clearly on the southeastern limb of the Llŷn Syncline. Cleavage (S_1) dips 80° towards 320°. The mudstones yield a Lower Longvillian shelly fauna

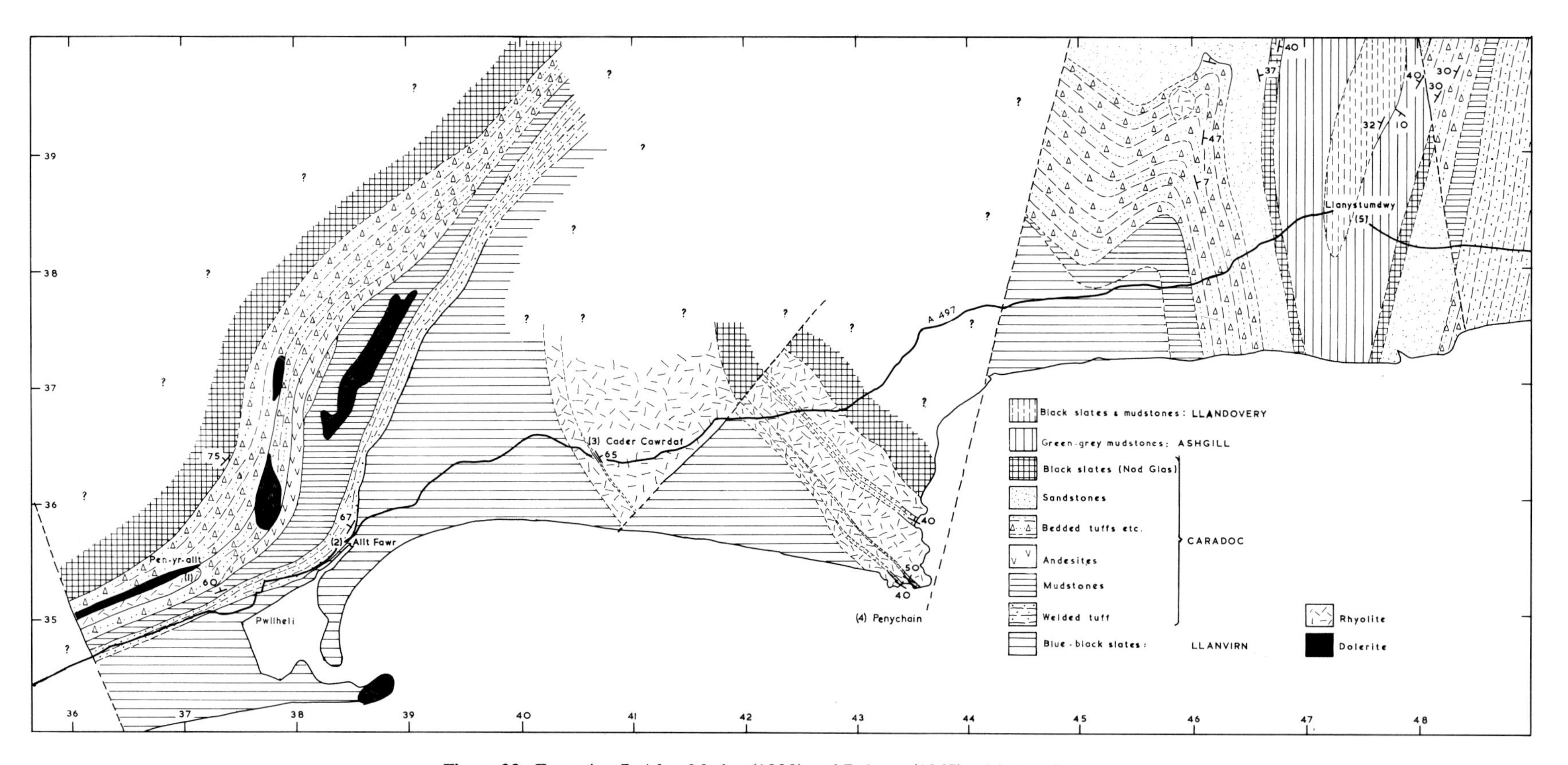

Figure 23. Excursion 7. After Matley (1938) and Roberts (1967), with permission.

but collecting is not advisable in this locality! The mudstones are overlain by deeply weathered, pale green, vesicular andesites which, in some places, may be pillowed but it is difficult to be certain. The lavas contain several horizons of blue, silty mudstone and exposures continue to the top of the hill.

Just beyond the school take the track leading southwest onto Pen-yr-allt. South of the trigonometrical point overlooking the town are exposures of a superb agglomeratic, andesitic tuff. The clasts are of rhyolite, andesite, basic pumice and sediments in a matrix of andesitic crystal-lithic tuff. Immediately south of the trigonometrical point is a small depression, and then flow-banded rhyolite is exposed. The rough, weathered surface and the regularity of the banding suggest, at first, that the rock might be a bedded rhyolitic tuff. It is overlain northward by about 5 m of green, eutaxitic welded tuff indicating temporary emergence and establishment of subaerial conditions. More flow-banded rhyolite succeeds the welded tuff. The rhyolite develops a nodular facies immediately north of the trig. point and flowage-folding is developed. The rhyolites are probably intrusive. A poorly exposed dolerite is present in the field to the north.

Return to the road and walk northwest to the turn off to the cemetery. Crudely bedded, andesitic tuffs and agglomerate dipping steeply northwest are exposed in the roadside. *Continue down the road to Pont y Ddwyryd where a stile on the right built into the wall gives access to the stream bed.* Here a conglomerate with a blue, sandy and muddy matrix overlying the volcanic horizon is exposed.

Return towards Pwllheli, but take the first turn on the left, and turn left again in 200 m. Walk north to the Afon Ddwyryd, cross it and follow the footpath to [373363]. Small exposures of sooty black slates of the Nod Glas sediments dipping 75° towards 310° overlie the conglomerate. The slates yield graptolites of the *D. clingani* zone. Notice that the strike has swung from 070° to 040°. The thin conglomerate and the succeeding black graptolitic slates indicate that the variable shallow-water and subaerial conditions which prevailed during the accumulation of the volcanic group gave way to quiet, marine conditions of deeper water wherein the black slates probably accumulated beneath the reach of wave base.

Return to the road that leads northeast past the cemetery. At [377358] examine exposures by a farm track on the right in strongly vesicular and flow-banded, grey-green andesite lava. About 100 m further up the road, at the turn off to Clogwyn Bach, are exposures of a coarse-grained, columnar-jointed dolerite sill emplaced within the volcanic succession.

Locality 2: Allt Fawr

Return to your transport and drive to [385357]. There are no parking problems.

A small roadside quarry (in which there is a transformer supported on a pair of poles) is cut in welded rhyolitic tuffs. The eutaxitic structure is difficult to pick up in the old quarry, but it is obvious on the weathered surfaces exposed on the hillside above the quarry. The eutaxitic structure dips 67° towards 300°, and the attitude of the columnar jointing suggests a similar dip for the sheet. Several ignimbrite flows are present separated by thin, non-welded, vitric tuff. These members of the volcanic group clearly accumulated under subaerial conditions.

Descend to the A499 road and take the narrow road leading northwest from [382356]. After 100 m a small quarry on the right is cut in sandstones with thin mudstones and siltstones. The sediments, which show parallel lamination, are blue-grey on fresh surfaces but weather brown. Brachiopods and trilobites of probable Lower Longvillian age are abundant. The sediments are on the same horizon as the mudstones seen at the start of the traverse of Locality 1 and thus the rhyolitic tuffs of Allt Fawr are the oldest Caradoc rocks of the succession.

Locality 3: Cader Cawrdaf

Take the Criccieth road as far as the road junction at [405365]. Walk east along the A497 for 120 m to a low cutting in front of a house called 'Corwel'.

A khaki-weathering mudstone about 1·5 m thick is intercalated with a feldspar-porphyritic, blue rock of andesitic appearance. The mudstone dips 65° towards 060° and hence an anticlinal axial trace lies to the west between this locality and Pwllheli. This structure has been referred to as the Abererch Anticline by Shackleton, and it clearly plunges northeast. Further east along the road are good exposures of orange-weathering, flow-banded rhyolite. The flow bands usually dip away to the northeast, but sometimes show intricate flowage-folding. They are here interpreted as intrusive rhyolites, whereas Tremlett (1965) has mapped them as bedded rhyolitic tuffs.

Locality 4: Penychain

Drive east for 2·5 km and turn south immediately before Butlin's holiday camp. Park on the verge north of Penychain house and follow the track to the lowest exposed rocks at [433353].

The lowest rocks in the succession are exposed on the westernmost point and consist of flow-banded rhyolite. The base of the sheet at the end of the beach is superbly flow-brecciated and flowage-folded. The overall banding and the attitude of the columnar jointing suggest a sheet dipping at about 50° towards about 045°. The rhyolite is presumed to be an intrusive sheet. It is overlain in the bay to the east by bedded agglomeratic tuffs and lapilli tuffs. The clasts are mainly of rhyolite in a dark blue, rhyolitic matrix. The bedding dips 40° towards 030°. The bedded tuffs and agglomeratic tuffs are overlain by a flow-banded, flowage-folded rhyolite which forms the crag with the trigonometrical point. The columnar jointing suggests the same dip and strike.

Just around Penychain point the 'trig. point rhyolite' is overlain by another agglomeratic tuff, about 4 m thick, through which have been intruded thin stringers of rhyolite. The dip of the bedding has steepened to 50° towards 030°. The clasts again are mainly of rhyolite, but the matrix is basic and green-weathering. Cleavage (S_1) dips at 85° towards 347°. This tuff is overlain by another flow-banded, flowage-folded rhyolite which has a very irregular base. It becomes columnar-jointed upward and then comes a strongly brecciated horizon with blocks up to 2 m across. Flowage-folding becomes obvious in coarsely banded rhyolite and then, immediately before the small beach with its 20 m exposure gap, there is a return to a massive, autobrecciated rhyolite.

Immediately north of the beach the rocks are, once again, finely flow-banded, flowage-folded rhyolites with a prominent nodular horizon some 2–3 m above the apparent base.

This is succeeded by coarsely banded and brecciated rhyolite, and the next headland is composed of finely flow-banded rhyolite.

The next inlet has a remarkably intense nodular development in flow-folded rhyolite and some of the rocks are very deeply weathered. The small headland to the north consists mainly of brecciated rhyolite and then there is a 110 m long bay.

Bedded rhyolitic and andesitic tuffs are exposed at the extreme southern end of the bay and dip 40° towards 030°. They are succeeded immediately by eutaxitic, welded tuffs of bright green, blue, white and grey which are rich in lapilli of rhyolite, pumice and fine-grained, black rock which may be pelite. Columnar jointing is obvious.

Two-thirds of the way across the beach there is a prominent crag in which an irregular junction of the agglomeratic, welded tuffs with the overlying flow-banded, flowage-folded and autobrecciated rhyolite may be examined. The flow-banded and folded rhyolite forms the final headland and exposures of solid rock cease against boulder clay.

The bulk of the rocks in the Penychain traverse thus show the field characteristics of rhyolites and are probably intrusive. As at Locality 3, Tremlett (1965) has interpreted them otherwise: as bedded tuffs. The presence of welded tuffs indicates that at least some of the eruptive rocks accumulated in a subaerial environment. Although direct faunal evidence is lacking, the rocks are regarded as lying on the same horizon as the Pwllheli volcanics.

Locality 5: Llanystumdwy

Drive to Llanystumdwy and park near the Lloyd George memorial. A footpath leads from the memorial along the east bank of the Afon Dwyfor. There are no exposures, other than drift, until you pass the fisherman's notice which reads 'Association Waters End'.

Grey-green, silty mudstones and slates with siltstone laminae are exposed on the eastern bank and in the river at [477394]. The beds dip at angles of up to 10° towards 220° and have yielded an Ashgillian shelly fauna to Roberts (1967). The beds lie on the eastern limb of the Llwyd Mawr Syncline, although not far east of the axial trace. Wade across the river, and to the north, along the bank, jet black, pyritous mudstones of Llandovery age are exposed, which contain abundant monograptids preserved in pyrite in high relief at [478394]. The beds dip generally to the west and southwest, but minor folds are present. Micrograded, grey, silty laminae are present and a sub-vertical cleavage strikes at 000–010°. The axial trace of the Llwyd Mawr Syncline is thought to trend at about 010° about 400 m away to the west.

Carry on to the valley side beneath Dynana at [481395]. Green-grey mudstones and silty mudstones contain a good Ashgillian fauna and dip at up to 40° to the northwest. Southeastwards along the valley side they are underlain by black mudstone containing an Ashgillian, shelly fauna and immediately to the east comes a succession of volcanics, the highest of which is a strongly calcified agglomeratic tuff. Veins of pink rhodocrosite are present. The volcanics dip at 30° towards 300°. The volcanics are exposed along the northwest bank upstream as far as [485398], after which the bank is of drift. Rock types exposed include welded rhyolitic tuffs, flinty vitric tuffs, crystal-vitric tuffs and bedded pumice lapilli tuffs. Dips are consistently northwest into the Llwyd Mawr syncline. The junction between the Caradocian volcanic rocks and the fossiliferous Ashgillian sediments is faulted and, as a consequence, the fossiliferous topmost Caradocian sediments, which are present above the volcanics in the western limb of the syncline, have been cut out here in the eastern limb.

Eifionydd

EXCURSION 8. THE YNYSCYNHAIARN ANTICLINE

The main aim of this excursion is to examine the Upper Cambrian succession exposed in the northerly plunging anticline. The sequence is tabulated in the Cambrian correlation chart in figure 4 (column 5), and the Ordovician correlation chart in figure 5 (column 5).

Criccieth is well supplied with car parks.
Use the excursion geological map, figure 24.
Reference: Fearnsides (1910).
1:50 000 OS sheet No 124; or 1:25 000 sheet No. SH53.

Locality 1: Criccieth Castle

The castle stands on a sill of intrusive rhyolite. Descend to the beach and examine the cliff exposures. The rock is an orange- to buff-weathering, flow-banded, flowage-folded, feldspar-porphyritic rhyolite. Fresh surfaces are blue or greenish grey. Columnar jointing is well developed on the eastern side of the intrusion, and again at Dinas, northwest of the castle. The jointing suggests that the mass is part of a sill-like sheet dipping westward, which has been emplaced in the western limb of the Ynyscynhaiarn Anticline.

Locality 2: [507381]

Walk eastward along the front and beyond the groyne to a low cliff cut in boulder clay.

The small intertidal point is due to the residual train of boulders remaining after the clay matrix has been washed away by wave action. The boulder train includes a magnificent assortment of wave-polished rock types, including welded tuffs, bedded tuffs, agglomerates, flow-banded rhyolites, dolerites, coarse gabbros, greywackes, etc.

The cliff shows 4 m of disturbed, buff-coloured till overlying 2 m of blue till.

Locality 3: Traverse of Rhiw-for-fawr

Continue eastward alongside the railway.

Note the small lagoon which is silting up behind the pebble storm beach. Examine the low crags at the eastern end of the lagoon at [512381]. They consist of blue-black, pyritous, silty slates, the silty laminae of which weather out. These are the Moelygest Beds of Fearnsides. Bedding dips 42° towards 288° whereas the slaty cleavage, S_1, dips at 80° towards 282°. The exposures are therefore in the western limb of the anticline and the bedding/cleavage intersection indicates a plunge of 6° towards 012°.

The beds become less silty as a low wall is approached, and east of the wall is a grass-covered depression about 8 m wide concealing the famous *Dictyonema* band. Rare slabs of the rock can be found; it is a faintly colour-laminated, dark blue-grey, pyritous slate which, when it splits along the bedding, is sometimes crowded with *Dictyonema*.

The eastern side of the depression is formed by a crag of interbedded blue, black and grey, pyritous siltstones, fine-grained sandstones, and

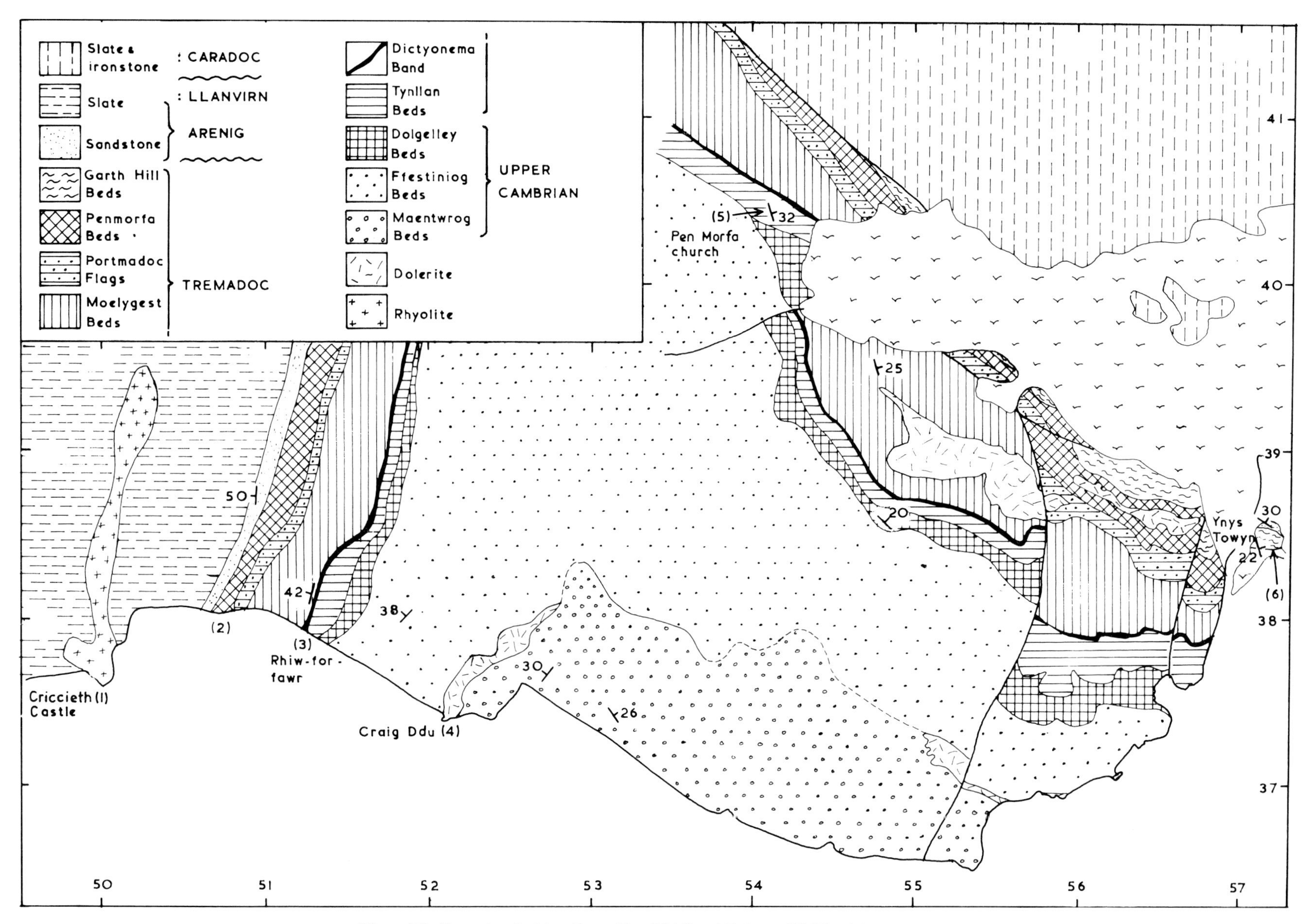

Figure 24. Excursion 8. After Fearnsides (1910) and Roberts (1967), with permission.

mudstones. The rocks are strongly cleaved but the coarser-grained types can be split fairly readily along the bedding to yield trilobites and small, horny brachiopods. These sediments constitute the Tynllan or *Niobe* Beds of Fearnsides, and extend eastward to the small railway cutting of Ogof ddu. Ogof ddu has been excavated along the course of three small sub-parallel faults of negligible throw. The Tynllan Beds are the lowest members of the Tremadoc Slates. The abundant blue-black pyritous mudstones and slates and the associated parallel-laminated siltstones suggest that accumulation of the Tynllan Beds, the *Dictyonema* band and the Moelygest Beds took place under quiet marine conditions beneath the reach of wave base. The occasional fine-grained sandstones are probably the result of turbidity current action.

The Tremadoc Slates are underlain by the Dolgelly Beds, which are dark blue-black rocks which form the cliff from Ogof ddu eastwards for 80 m to a prominent, low, grey crag in the field north of the railway line. At Ogof ddu the highest Dolgelly Beds are blocky mudstones. They are underlain by a series of finely laminated, blue-black, pyritous slates, mudstones, siltstones, and thin, fine-grained sandstones. Pass through the wire fence alongside the railway into the old quarry, where a 25 cm thick, yellow, feldspathic tuff is very prominent low down in the quarry face amongst the blue-black, muddy sediments. Trilobites and small brachiopods are easily obtained from the blue-black slates overlying the tuff. Beneath the tuff are 4 m of laminated slates and sandstones. Large, ellipsoidal, sideritic concretions (1 m × 60 cm) have grown in the sandy horizons, and a cone-in-cone structure has developed in the muddy sediment at the margins. About 20 m of black mudstones and slates underlie the concretionary beds and the oldest Dolgelly Beds consist of about 10 m of dark grey and black siltstones and very fine-grained sandstones. The Dolgelly Beds clearly accumulated under conditions similar to those which obtained during the deposition of the overlying Tremadoc sediments.

The underlying Ffestiniog Beds begin with grey siltstones and fine-grained sandstones with *Lingulella davisii* (McCoy). Occasional bedding planes are crowded with the brachiopod and this horizon constitutes the *Lingulella* band. The horizon is underlain by a sandstone which shows a 'box weathering' rather like some ironstones.

Continue along the old sea cliff past a second lagoon where Ffestiniog quartz-rich greywacke sandstones with thin interbedded shales show a variety of sedimentary structures, including cross lamination, ripples, load casts, flute casts and scours. About 20 m beyond a fence at the end of the lagoon, shell banks 10–15 cm thick, consisting of innumerable *Lingulella davisii*, are developed.

Locality 4: Craig Ddu

Cross the railway line at Blackrock Halt and walk to the old quarries in Ffestiniog Beds at [521376].

The rocks exposed are near the base of the Ffestiniog Beds and consist of lenses and scour fillings of ripple-drifted, fine-grained sandstones, interlaminated with siltstones and silty mudstones (figure 25). Load casts have developed on the base of some of the sands. The Ffestiniog Beds, with their abundant evidence of current action, provide a strong contrast with the overlying Dolgelly Beds. Some of the sediments have charac-

Figure 25. Flaser bedding in Ffestiniog Beds near Blackrock Halt (Craig Ddu). Lenses and scour fillings of cross laminated fine sands and silts within siltstones and silty mudstones.

teristics of turbidites, but the material of the shell banks has not travelled far because, although the *Lingulella* valves are disarticulated, they remain unbroken. Crimes (1970a) believed the sediments to be sub-littoral and to have been deposited from bottom currents less spasmodic than turbidity currents.

The ripple-drifted beds immediately overlie a dolerite sill that forms the hill of Craig Ddu. Examine the top of the sill and notice that the fine-grained marginal facies is cleaved, although the interior is coarser-grained and apparently uncleaved. Good columnar jointing is developed. The sill was evidently intruded before the main phase of folding.

The Maentwrog Beds are exposed beneath the sill and they can be examined along the shore if the tide is out or along the cliff if it is not. They consist of parallel-laminated, greywacke siltstones and fine-grained sandstones 5–15 cm in thickness. The greywackes show a wealth of sole markings, including groove and flute casts, as well as load casts, convolute lamination and graded bedding.

Figure 26. Minor folds in Maentwrog sandstones, siltstones and slates at Craig Ddu. Note the good axial-planar cleavage, the small fault to the right of the hammer and the carious weathering in the sandy beds due to a patchy distribution of cement.

At the start of Blackrock Sands, and rather lower in the succession, the beds become more pelitic and the greywacke sandstones are finer-grained and thinner. If the tide has compelled you to keep to the cliff top path you can descend to the beach via a thin dolerite dyke trending at 070°. Having gained the beach, the cliff section shows a series of northerly plunging minor folds with well developed axial-planar cleavage, S_1 (figure 26).

The greywacke sandstones and siltstones show characters of turbidites and the succession indicates that turbidity current action became more important towards the top of the sequence.

Follow the old sea cliff to about [528377], where the hinge of the Ynyscynhaiarn Anticline can be located.

Locality 5: Pen Morfa Church

Return to the transport and take the road towards Porthmadog.

East of Pentre'r-felin, roadside exposures of Ffestiniog greywackes and shales are worthy of study, if time allows. Again, an extensive new road cutting at [546396] in blue, silty mudstones with thin greywacke sandstones of the Moelygest Beds yields occasional brachiopods and trilobites. There is a large, convenient lay-by in which to park.

Continue west through Pen Morfa to [540407] where there is a lay-by on the south side of the road. Take the track leading southeast towards Pen Morfa church. At the former bridge which carried the now dismantled railway over the track, leave the track and follow the old railway east for about 30 m.

A small depression running down the field marks the position of the *Dictyonema* band. Its outcrop in the old railway cutting is largely overgrown and, at the time of writing, is guarded by five hives of bees; but specimens carrying *Dictyonema* can be found. (Keep out of the line of flight of the bees.)

Return to the track and continue southeast. 120 m north of the end of the track, the *Niobe* Beds are exposed in a crag in the field on the west and in the stream. Trilobites are fairly readily obtained.

Continue south to the minor road and turn east to Pen Morfa village. At the road junction in the village are small exposures of the Penmorfa Beds. These are blue-black, silty slates which, in the past, have yielded a rich trilobite fauna. The Tremadoc succession is terminated, at the chapel by the last bend in the road, by the Penmorfa Fault which Fearnsides thought was a thrust.

A short distance southeast of Pen Morfa, road widening has produced a fresh exposure of the overlying Tyddyn-dicwm Beds of the Caradoc in which blue-black slates with chamositic ooliths and nodules of chert are seen. Graptolites are common and include *Nemagraptus gracilis* (Hall) and *Climacograptus peltifer* (Lapworth). Pisolitic and oolitic ironstones are exposed in a low ridge immediately northeast of the road.

Locality 6: Ynys Towyn

There is usually space to park in the small square off the main road.

The first exposures are the Garth Hill Beds which consist of 30 m of rusty-weathering slate, silts and fine-grained sandstones. Beautifully distorted specimens of *Angelina sedgwicki* (Salter) can usually be turned out here. Beds dip 22° towards 076°. Continue past the Gwynedd River Authority building, behind which the massive, basal Arenig Garth Grit can be seen resting on the Tremadoc.

The basal bed rests on an irregular erosion surface. The Garth Grit is about 20 m thick and consists of coarse, quartz-rich sandstones and conglomerates cemented with silica. Clasts are mainly of quartz but some mud-pellet-rich horizons are present. The beds dip at 30° towards 032° and contain the chemogenic nodule formerly called *B. undosa*. The conglomeratic Garth Grit of Arenig age therefore rests unconformably upon the underlying Garth Hill Beds of the Tremadoc and the relationships indicate that the Garth Grit can be expected to cut down onto older members of the Tremadoc to the northwest.

EXCURSION 9. LLWYD MAWR

The purpose of this excursion is to examine the Llwyd Mawr ignimbrite sheet with its associated volcanic rocks, and to examine the deformation in the Cwm Pennant Anticline. The sequence is tabulated in the Cambrian correlation chart in figure 4 (column 6); and the Ordovician correlation chart in figure 5 (column 4).

Use the excursion geological map, figure 27.
References: Roberts (1967, 1969, 1975), Roberts and Siddans (1971), Shackleton (1959) and Wright (1974).
1:50 000 OS sheet No. 115; or 1:25 000 sheets Nos SH44, 45, 54 and 55.
The excursion involves a pleasant day's hill walking, a little scrambling, and a round trip of about 12–15 km. A car can usually be left for the day at the end of the metalled road at [496511].

Locality 1: Cwm Silin

Take the track which leads first east, and then southeast into Cwm Silin. Until Llynau Cwm Silin are reached, the solid rock (Llanvirn slates) is concealed beneath a thick cover of morainic debris. Make for the start of the western wall of the cwm at [511504], where a small stream rises.

Soft blue-black and grey-black slates with rare silty laminae are seen overlain by strongly cleaved, unbedded, rhyolitic, pumice-lapilli tuff. The slates yield rare graptolites and were deposited under quiet marine conditions beneath the reach of wave base. The cleaved tuff represents the basal, non-welded facies of the subaerial Llwyd Mawr ignimbrite. The rock is deeply weathered and sufficiently coarse-grained and recrystalline to be termed a phyllite. It is a grey, buff-weathering, fissile rock composed of a base of quartz, orientated grey-blue micas, and rotted, buff and cream feldspars; the lapilli are flattened in the plane of cleavage. In thin section it can be seen to be a stilpnomelane–muscovite–quartz phyllite.

If the correlation of the Llwyd Mawr ignimbrite with the Pitt's Head ignimbrites is accepted, then the Llwyd Mawr ignimbrite rests unconformably on the Llanvirn slates because the Llandeilo and the Glanrafon Beds of the Caradoc are missing.

Make your way along the wall of the cwm to the foot of the exposures above the scree. About 5 m above the base, pumice lapilli and occasional

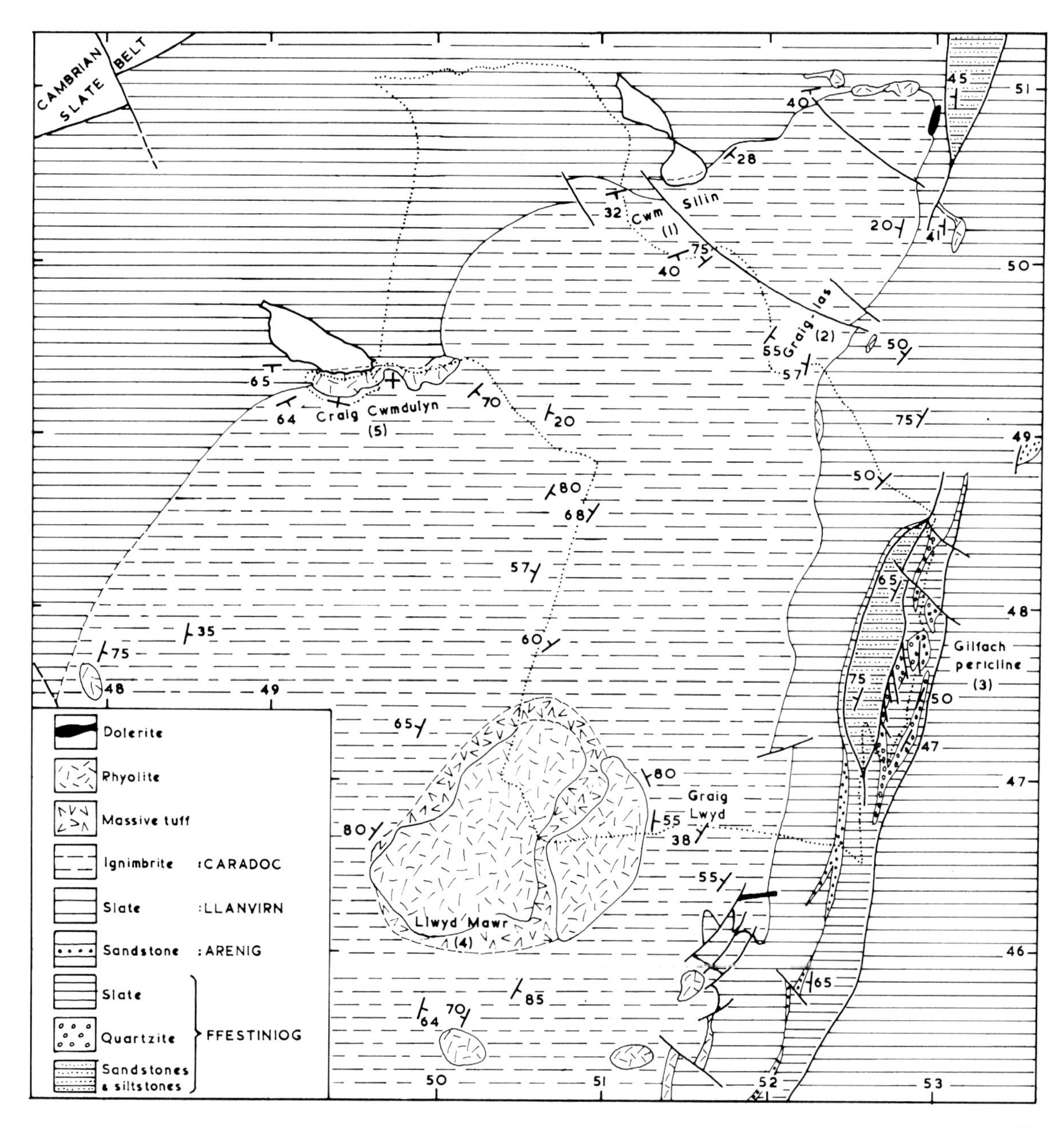

Figure 27. Excursion 9. After Roberts (1967) and Crimes (1969b), with permission.

rhyolite blocks become more obvious, but for the next 15 m the tuff is still very strongly cleaved. The rock then becomes more compact and resistant, and about 20 m above the base of the sheet, although still obviously strongly cleaved, the pumice lapilli are now flattened in a plane inclined at 35° to the southsoutheast. The flattened lapilli define a eutaxitic structure which is parallel to the base of the sheet, and has the same structural significance as bedding in bedded rocks. Columnar jointing is now obvious and becomes perfectly developed in many places higher in the sheet.

Continue the traverse towards the prominent cleft at the head of the cwm and south of the buttress of Craig Cwm Silin. The pumice lapilli approximate to ellipsoids (fiamme), which show increased deformation (due to welding and compaction) with height above the base of the ignimbrite sheet, and the rock takes on a blue-grey, flinty appearance on freshly fractured surfaces. At a distance of 500 m from the start of the traverse a synclinal axial trace is crossed and the eutaxitic structure now dips northwest. The tuff is now intensely welded, and fiamme show such intense flattening that a small hand specimen might be mistaken for a flow-banded rhyolite.

Scramble up through the broad cleft at the head of the cwm. At the top of the cleft notice that an anticlinal axial trace is crossed and that the eutaxitic structure returns to a southeasterly dip. The traverse has therefore crossed a minor fold-pair within the Llwyd Mawr Syncline.

Locality 2: Graig-las

Traverse southeast across the summit plateau to the cliffs of Graig-las. (There is a convenient wall you can follow if the cloud is down and you are navigating by time and compass.)

The cliff is in two tiers. A minor synclinal trace runs close to the base of the upper tier and crosses the wall where it runs south for about 70 m. The ignimbrite is intensely welded and massively columnar-jointed. As you descend the crag and the succession, the columnar jointing becomes closer as the degree of welding becomes less intense until, about 50 m above the base of the sheet, the jointing becomes platy in a feebly welded tuff. The base of the sheet consists of some 10–30 m of non-welded, unbedded, pumice-lapilli tuff. The cliffs provide readily accessible exposures through a thickness of about 350 m, where the variation in welding (as indicated by the deformation of the pumice lapilli) and in the jointing pattern with height above the base of the sheet, can be conveniently studied.

Locality 3: The Gilfach Pericline

Walk southeast across sporadic exposures of soft, blue-black Llanvirn slate to [530486] where the stream immediately south of Nant Braich-y-ddinas turns southwest through 90°. In 50 m the stream turns southeast through another right angle and enters a gorge. Keep to the south of the gorge.

Examine the Pennant Quartzite cut off at the gorge by a fault. It is a massive, pebbly quartzite here lying in the western limb of the Cwm Pennant Anticline. The axial trace of the anticline lies just to the east of the quartzite and can be located in the underlying tough, grey-black, parallel-laminated, silty slates.

Follow the anticlinal axial trace southsouthwest for 300 m, where it is lost against a northwesterly trending fault which has shifted the axial trace 50 m to the east. South of the fault, which evidently has a sinistral tear component, the quartzite occurs as a series of pods surrounded by slates as a result of tectonic slicing and boudinage in the tight core of the anticline. Continue southsouthwest alongside Ceunant Ciprwth, where it flows along the axial trace of the anticline. The stream turns southeast and cuts through the Pennant Quartzite which forms the southeastern limb of the fold; 250 m southsouthwest of this point the southern closure of the pericline is seen as the limbs of quartzite converge at [526471]. Roberts (1967) regarded this quartzite and the associated slates as of Ffestiniog age.

Walk northwest to prominent crags of a series of fine-grained sandstones, siltstones, thin silty slates and slates. The sandstones show convolute lamination, cross bedding, ripple-drift lamination and flute casts. The sediments probably accumulated in a sub-littoral environment. Shackleton (1959) and Roberts (1967) both regarded these rocks as Ffestiniog. The rocks constitute a fault slice in the western limb of the Cwm Pennant Anticline.

Walk south to an old copper mine and prominent exposures at [524467]. Tough, splintery, olive-grey slates with psammitic laminae which overlie the Pennant Quartzite (previously seen in the pericline and regarded as Ffestiniog in age), are overlain by a conglomeratic greywacke sandstone. Roberts (1967) tentatively grouped this conglomeratic greywacke (which is the Upper Pennant Quartzite of Shackleton 1959) with the overlying Llanvirn slates. Crimes (1969b) has shown it to be more probably Arenig in age because it contains *Phycodes circinatum*.

Locality 4: Llwyd Mawr

Walk west across sporadic exposures of Llanvirn slates to the base of the ignimbrite sheet. A non-welded, strongly cleaved, unbedded, rhyolitic, pumice-lapilli tuff once again overlies the slates. It is about 10 m thick and passes upward into eutaxitic, welded tuff. Continue to Graig Lwyd where, at higher horizons in the sheet, an intensely welded tuff is exposed. A minor synclinal axial trace is then crossed.

Continue on a bearing of 255° across a poorly exposed, flow-banded, flow-folded, intrusive rhyolite to [507467] where, near the wall, two masses of flow-banded and flowage-folded rhyolite, with a nodular marginal facies on the eastern rhyolite, are separated by about 50 m of khaki-weathering, crystal-vitric tuff. The tuff is non-welded, unbedded, strongly cleaved and rich in feldspars. It is interpreted as a vitro-clastic tuff which filled a vent drilled through the ignimbrite sheet.

Walk northwest for 250 m to prominent exposures of flow-banded and flowage-folded rhyolite. Note the apparently random plunges of the flowage folds.

Locality 5: Craig Cwmdulyn

Follow the footpath which keeps to the west side of the wall and leads northnortheast over Llwyd Mawr for 2 km to Bwlch Cwmdulyn. At the bwlch turn northwest down the pass.

Continue for 1 km down the pass across the strike and through a thickness of about 750 m of welded tuffs dipping predominantly southeast. The pass turns west and immediately the roof of an intrusive dome is exposed in the stream and south of the stream. The rock is beautifully flow-banded and autobrecciated. Continue across the peat-covered flat to the crag at the south side of the head of Llyn Cwmdulyn, where a rhyolite dome is exposed in vertical section. The interior of the dome is flow-banded and flowage-folded, but the marginal facies (about 10 m thick) consists of autobrecciated rhyolite. The junction is against ignimbrite, but two lenses of blue-black Llanvirn slate are caught up along the junction.

Scramble up the contact and across the top of the dome and begin to descend the western contact. Look up and observe the smashed and bent, columnar-jointed, welded tuff against the margin of the dome. Wright (1974) has suggested that this rhyolite dome was extrusive and pre-dated the emplacement of the Llwyd Mawr ignimbrite sheet. Roberts (1975) continued to prefer an intrusive origin for the rhyolite dome.

Cross to the northeastern shore of Llyn Cwmdulyn and look back at the Cwm Dulyn rhyolite dome. Observe the massive, crudely radiating, columnar joints.

Return to your transport across the rough pasture of Rhos-las.

Arfon

EXCURSION 10. CLYNNOG FAWR TO LLANAELHAEARN

Precambrian and Ordovician sedimentary and volcanic rocks and the small plutonic intrusions are examined. The excursion starts at Clynnog and ends at Llanaelhaearn: it can be quite a long walk back.
Use the excursion geological map, figure 28.
Reference: Tremlett (1964).
1:50 000 OS sheet No. 123; or 1:25 000 sheets Nos SH34, 44 and 45.

Locality 1: Cilcoed [425502]

The exposures of the Arvonian volcanic rocks are on enclosed land and crop out in a northwest-trending ridge on the hillside northeast of Cilcoed. Permission is necessary to enter the ground.

The rocks are mainly ignimbrites, being welded, rhyolitic tuffs with magnificent eutaxitic texture in thin section. Fiamme are uncommon; hence a eutaxitic *structure* is not often evident in the hand specimen. Several ignimbrite flows are present, and are separated in one exposure by a bedded rhyolitic tuff which dips 55° towards 155°.

Southeast of Cilcoed the volcanics are overlain by a conglomerate, probably representing the Cilgwyn Conglomerate. Sub-rounded clasts of rhyolitic volcanics up to 15 cm across are set in a matrix of quartz grains and tuffaceous debris.

Locality 2: Pen-y-garreg [425497]

Drive to the right-angled bend on the road at Pen-y-garreg, about 1 km east of Clynnog Fawr.

At the bend, an aplite sill is exposed. It can be seen to be flow-banded and flowage-folded in small exposures to the east along the northern flank of the ridge. The aplite is overlain to the south by a sequence of blue-grey, laminated, fine-grained sandstones, siltstones and slates. The sediments, which form the south slope of the prominent east–west-trending ridge, are exposed for about 100 m along the roadside. The beds dip at about 40° towards 155°, are they and the aplite are probably separated from the Arvonian volcanic rocks of Locality 1 by a strike fault. The age of the sediments is uncertain, since no fossils have yet been obtained from them. However, they are succeeded to the south by slates which have yielded *hirundo* zone graptolites to Tremlett (1964), and so the Pen-y-garreg sediments probably belong to the *extensus* zone of the Arenig.

Locality 3: Bwlch Mawr

Continue south to the road and take the track onto the northern slope of Bwlch Mawr. At the end of the track continue west to an old slate trial at [427487].

Greyish black slates have yielded graptolites indicating the zone of *D. hirundo.*

Ascend the hill a little to avoid walls, and then continue around to old slate workings at [433471]. The beds have been disturbed by soil creep and, although bedding can be seen, it is not possible to obtain a reliable measurement. The slates are blue-grey, sometimes silty, and contain worm borings and common biserial and pendant graptolites indicating the zone of *D. bifidus.*

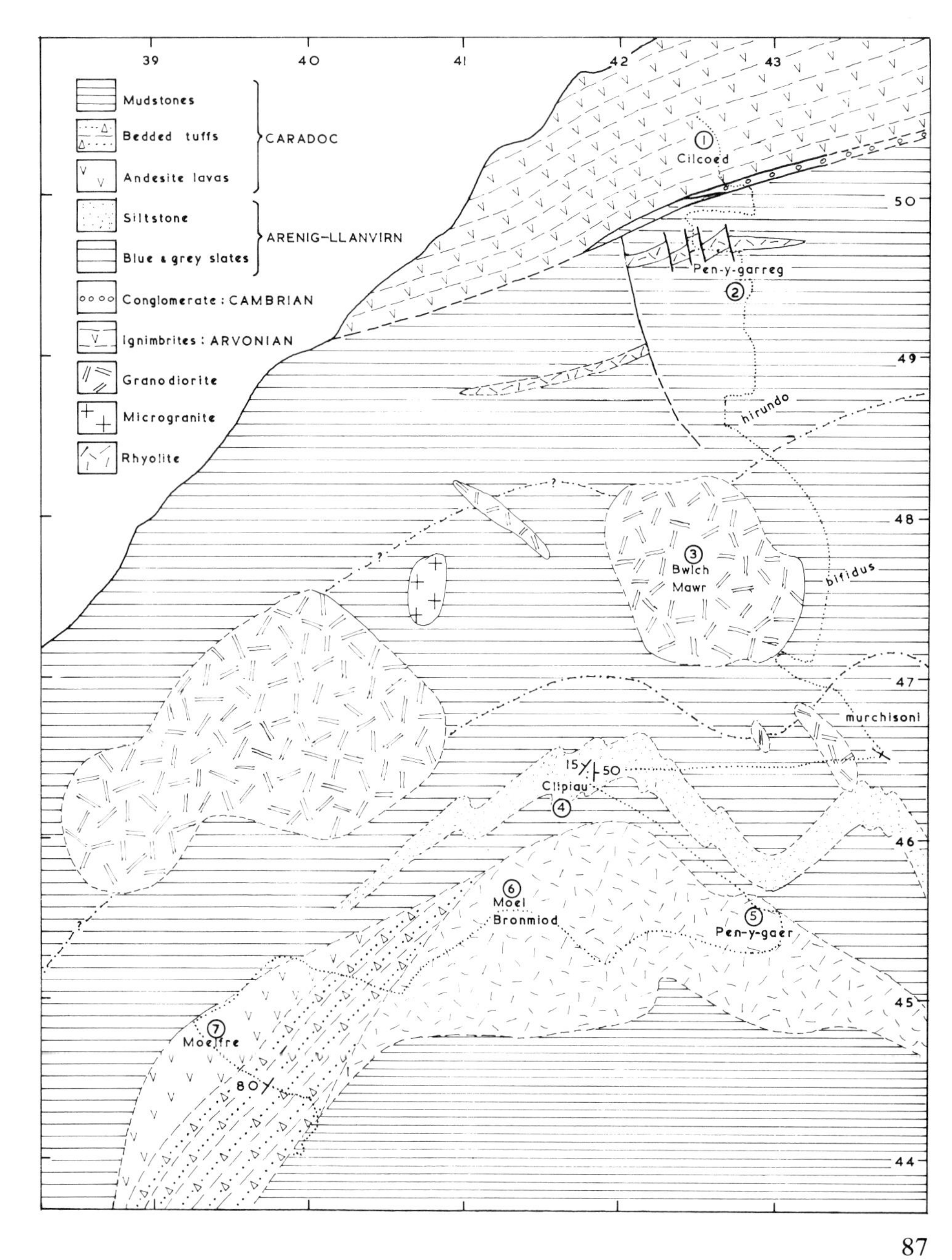

Figure 28. Excursion 10. Modified from Tremlett (1964), with permission.

Cross to the crags of Bwlch Mawr which are composed of a feldspar-porphyritic microgranodiorite. On a fresh surface the fine-grained matrix is pinkish with irregular green patches, apparently due to fine-grained chlorite. At this locality the jointing gives rise to long rectangular blocks, the long axes of which are inclined steeply to the east. The intrusion maps as a boss or plug.

Strike diagonally southeastwards down the hill to a small quarry on the west side of the road at [436466] where blue-black slates have yielded graptolites indicating the zone of *D. murchisoni.* Thin, silty laminae indicate that the bedding strikes at 295° and dips vertically. On the south side of the quarry the slates are cut by a 20 cm thick dyke of feldspar-porphyritic felsite. The dyke may be related to the Bwlch Mawr intrusion.

Locality 4: Clipiau

Walk west for 2 km over the hillside to the exposures at Clipiau at [417464] and [422465].

The rocks exposed are khaki-weathering siltstones with thin, fine-grained sandstones. Some of these are finely laminated, weather with a white porcellanous crust, have a conchoidal fracture, and may be rhyolitic vitric tuffs. No fossils have been obtained, but mapping suggests they lie within the *murchisoni* zone of the Llanvirn. The rocks have been affected by two phases of folding. F_1 minor folds trend between 0° and 035° (due to refolding about F_3 axes), are asymmetrical, and verge eastwards. An axial-planar, close fracture cleavage in the siltstone dips west at 55°. In areas where F_3 folds are absent, F_1 folds plunge at 5° towards 013°, but in many areas of the exposures, bedding and S_1 have been deformed by mesoscopic F_3 folds which plunge either at shallow angles due west or steeply due east, depending upon whether they have developed on a shallowly dipping western F_1 limb or a steeply dipping eastern F_1 limb. An axial-planar fracture cleavage, S_3, is present dipping at 80° towards 0°. It cuts the siltstones into pencil-like fragments.

Locality 5: Pen-y-Gaer [428455]

Walk southeast onto Pen-y-Gaer.

The rocks are shown on Tremlett's map as consisting of soda-rhyolitic types and are described as ignimbrites. In the field, the rocks have more the appearance of thick, brecciated rhyolites and over wide areas the banding dips 30° towards 240°. Nowhere was any trace of a eutaxitic structure found in the field nor, subsequently, when the rocks were examined in thin section. At the base of the summit crag on the southeast side of the hill, near to two obvious sheepfolds, the rocks show prominent flow-banding and flowage-folding. They are therefore interpreted as flow-banded and flowage-folded rhyolites and are probably of Caradocian age.

Locality 6: Moel Bronmiod [413456]

Walk west across the depression to the lowest exposures on the southeastern spur of Moel Bronmiod.

These rocks were referred to by Tremlett as soda-rhyolitic rocks and described as ignimbrites. Beautiful sweeping folds, plunging all over the

Figure 29. Crudely banded and brecciated rhyolite exposed on the southwestern slope of Moel Bronmiod.

place, can be seen and they are therefore interpreted as flow-banded and flowage-folded rhyolites. It is not known whether they represent flows or intrusions. Closely similar rocks are exposed at intervals to the summit, the only difference being that autobrecciation is sometimes well developed. Breccias and nodular structures are well developed at the summit itself and the crude layering resembles that of some agglomerates. However, the rocks are monomict: clasts are of rhyolite held in a rhyolitic matrix. They pass upward into flow-banded and flowage-folded, nodular rhyolite and, consequently, the summit breccias are interpreted here as most probably autobreccias.

In descending the hill towards the southwest, some very crudely banded and brecciated rhyolites are crossed (figure 29) and it may well be that these rocks consisting of almost sub-rounded rhyolitic blocks in a sparse, microcrystalline matrix represent rhyolite lavas. There are excellent exposures along each side of the wall that runs southwest down from the summit.

Locality 7: Moelfre [395447]

Continue southwest to the track and follow the road towards Moelfre.

Climb up the northwest-facing slope over a few small exposures and a small, old quarry in green, autobrecciated and flow-banded, andesite lava. The rocks are vesicular and contain autoclasts up to 50 cm long. The summit and the west-facing slope show magnificent exposures of purple-weathering, blocky and rubbly, andesite lavas. They are vesicular and amygdaloidal, vesicles being infilled with chlorite, chalcedony or carbonate. There is a great deal of jasper which fills amygdales and spaces between blocks, and which also occurs as regular and irregular veins. Walk southeast over the rough ground towards Moelfre-bach [399444]. Ignoring the huge erratic, the first exposure reached consists of a rubbly and blocky rhyolite with blocks up to 1 m across—perhaps a lava, possibly an intrusion—much veined and impregnated with jasper. It is not possible to determine the attitude of the rhyolite. The lowest exposure, where the track from the barn ends, consists of red, sandy, crystal-lithic tuffs, muddy tuffaceous sediments and crystal-vitric tuffs. Bedding dips at 80° towards 307°; cleavage, S_1, strikes at 052° and is vertical. The beds are the right way up and presumably folded on a minor scale to account for the northwesterly dip.

Note for the Reader Consulting Tremlett (1964)

Tremlett (1964) has misplaced Moelfre, together with the associated geology, by 1 km to the west on his map (Plate II). This error in drawing cannot really be said to detract seriously from the geology represented by his map.

EXCURSION 11. BANGOR TO CAERNARFON

The purpose of this excursion is to look at the geology of the Bangor Ridge. Arvonian, Cambrian, Ordovician and Carboniferous rocks will be examined.

Transport is essential if the ground is to be covered in one day. Even so, it may well be that most geologists would prefer to spend two days on such interesting, though not too well exposed ground. No ardous walking is involved. The excursion begins in Bangor and ends in Caernarfon.

Use the excursion geological map, figure 30.

References: Elles (1904), Greenly (1928, 1945) and Wood (1969).
1:50 000 OS sheet No. 115; or 1:25 000 sheets Nos SH46, 56 and 57.

Locality 1: Bangor Mountain and Castle Hill [584721]

Sandstones, siltstones and tuffaceous sandstones, often stained red and purple, are exposed on the northwest-facing slope of Bangor Mountain.

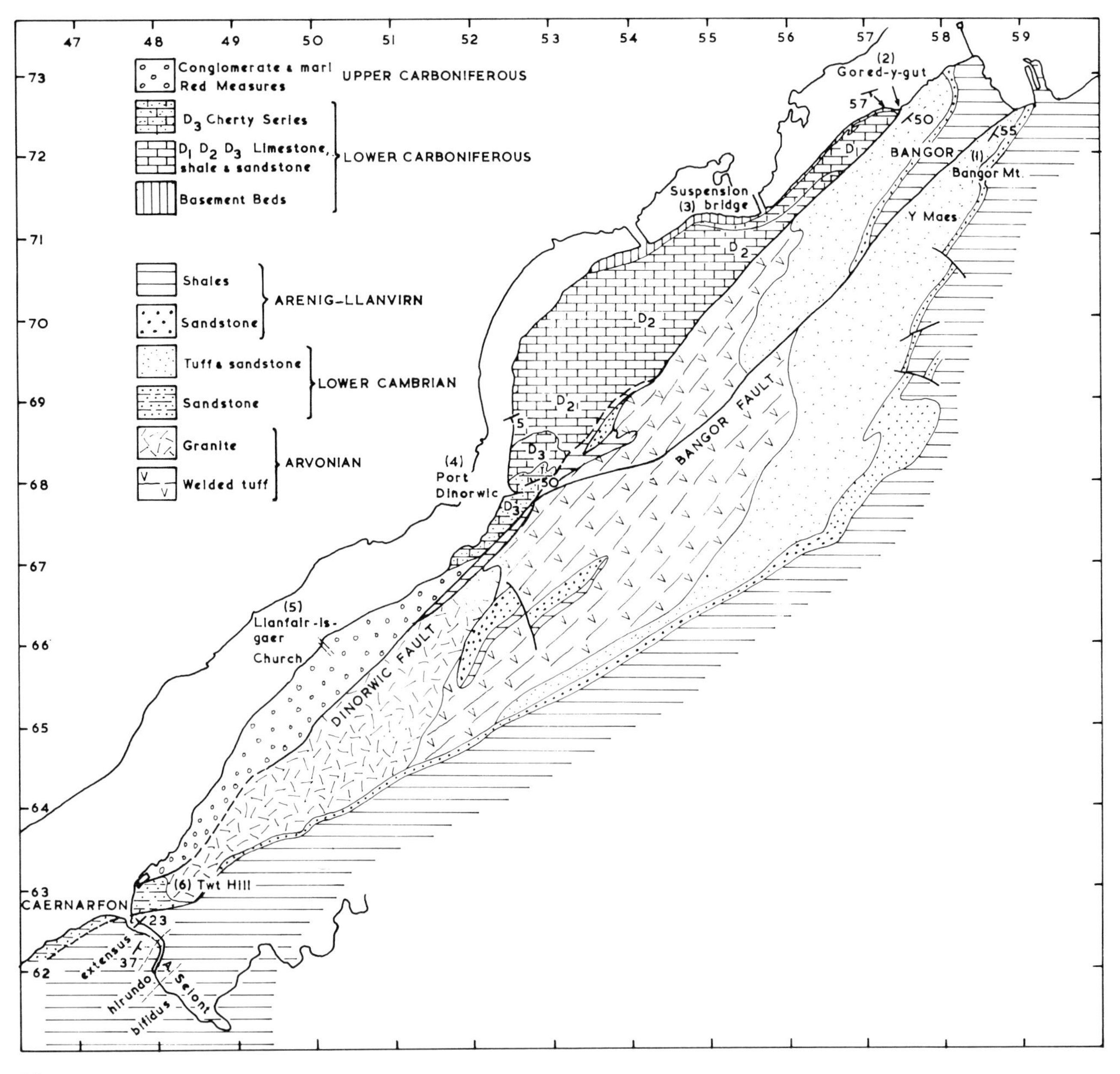

Figure 30. Excursion 11. After Elles (1904), Greenly (1928, 1945) and Wood (1969), with permission.

Beds dip at angles up to 55° towards 125°. Near Y Maes, conglomerates and sandstones are present; the succession was interpreted by Greenly (1945) as consisting of Arvonian tuffs which were supposed to occupy a syncline with a strong vergence to the northwest and which had been thrust northwestward over the Cambrian sandstones and conglomerates. Wood (1969) has shown that the tuffs and tuffaceous sediments are in normal stratigraphical sequence with the sandstones and conglomerates, and that all the rocks are therefore better regarded as Lower Cambrian.

The view northwest from Bangor Mountain is instructive. Bangor Mountain is bounded on the northwest by the Bangor Fault, which runs more or less parallel to the course of the High Street, and which downthrows to the northwest. Much of the city is built on the low-lying, down-faulted tract of Arenig shales, but Upper Bangor and the University College is sited on a ridge formed by the resistant Cambrian tuffs, sandstones and conglomerates which re-emerge, still dipping southeast, from beneath the overlying Arenig shales.

Locality 2: Gored-y-gut

From Upper Bangor, walk down the road to the foreshore of the straits at the former Youth Hostel at Gored-y-gut [574726].

Just before reaching the foreshore, note the exposures of conglomerate and sandstone on the southwest side of the road. The beds dip 50° towards 140°. They were regarded as Cambrian by Greenly (1945) and were thought to rest unconformably upon Arvonian rocks.

On reaching the foreshore, turn northeast to crags of grey-green sandstone. These are the 'viriditic tuffs' of Greenly. They contain pebble horizons and, in places, are parallel-laminated, which indicate that the rocks are slightly flexed but that generally they dip 50° to the southeast. The pebbles are of rhyolite and quartzite and the rocks are cut by shears and quartz veins, probably related to the adjacent Dinorwic Fault. The grey-green sandstones are overlain by massive, crudely bedded conglomerates and sandstones with the same dip and strike. This junction was interpreted as an unconformity by Greenly.

Walk southwest along the foreshore, across the trace of the Dinorwic Fault, onto the Lower Carboniferous Basement Series exposed near the former Youth Hostel. A series of sandstones, shales, conglomerates, calcareous sandstones and limestones are exposed dipping 57° towards 165°. The beds are wedge-shaped rather than tabular, and the sandstones have irregular erosional bases. Cross bedding is usual. The pebbles in the conglomerates are mostly quartzite and jasper. The sandstones carry worm burrows, the shales are dark and carbonaceous and contain fragmentary plant remains. The limestones are up to 25 cm thick; some are nodular, whereas others are brown, porcellanous, micritic limestones. The beds are marine and were laid down in shallow water affected by strong currents; probably in a sub-littoral environment. The terrigenous clastic component is not far-travelled and was derived mainly from the Mona Complex.

Continue southwest for 400 m to exposures in the low cliff of a limestone/shale sequence still within the Basement Series. The limestones are bioclastic, rich in bits of crinoids and brachiopods, and are up to 30 cm thick. Some limestone beds have rippled and irregular bases but all have flat tops. The shales are well laminated, brown, sandy and calcareous. Deepening of the water and some lessening of current activity is suggested. The fauna indicates the beds to be of D_1 age.

At the small point at [568724] a spheroidally weathered, Tertiary dolerite dyke, trending at 280°, cuts iron-rich, quartz-granule sandstones interbedded with siltstones, laminated mudstones and lenticular beds of limestone and calcareous sandstone.

At a point just east of Cae Coch, a house on the shore, climb up the grassy cliff to an old quarry in a wood. The quarry is cut in limestones, shales and sandstones of the D_1 zone. Beds of sandstone and shale up to 2 m thick make up the lower part of the succession. Climb up the slope at the eastern end of the long quarry to the irregularly bedded limestones at the top of the exposures. The limestones are bioclastic and are rich in crinoid remains and large productids.

Locality 3: Suspension Bridge

There is usually enough room to park a car on the side of the road leading to the University Sports Ground [557712]. The foreshore can be reached from the west side of the bridge, east of the grounds of the Cerris Nursing Home.

The Basement Series exposed immediately west of the bridge consists of quartz-pebble conglomerates, sandstones and shales overlain by distinctive red and green, silty and sandy marls. About 220 m to the west, just past the house on the foreshore, Arenig shales are exposed dipping 30° towards 110°. The shales, which contain graptolites, including *D. extensus* (Hall), are stained red and purple, but the staining is superficial and the rocks are blue-black when broken across. The Basement Series can be seen here resting unconformably upon an irregular surface cut in the Arenig shales. The lowest bed of the Basement Series here is a red-purple, pisolitic ironstone in a very friable condition. It contains dispersed, sub-angular and sub-rounded clasts of Mona Complex schists and quartzites which range from sand- to boulder-grade, as well as small chips of Arenig shale. It is the 'Loam-Breccia' formation of Greenly.

Locality 4: Porth Dinorwic

Drive to Porth Dinorwic. There is plenty of space to park a car clear of the road near the dock [527678].

Cross the dock and examine the beds exposed along the north side of the dock. They are of D_3 age and consist of cherty and rubbly, crinoidal limestones with interbedded, thin, calcareous sandstones. Bedding dips steeply northwest and has probably been turned into this attitude against the Dinorwic Fault which runs, with a Caledonoid trend, about 120 m away to the southeast.

Take the road leading northwest towards the new housing estate. A roadside cutting beneath Plas Dinorwic exposes a small anticlinal hinge in thinly bedded, quartz-pebble conglomerates, thin rubbly limestones, cherts and sandstones which lie above the crinoidal limestones and sandstones seen at the dockside. The Plas Dinorwic Beds constitute 'the Cherty Series' of Greenly and they are preserved in a synclinal structure plunging gently southwest, within which the anticlinal hinge is a minor flexure.

Continue through the housing estate and onto the foreshore where pink-weathering, fine-grained, rubbly limestones of D_3 age with *Lithostrotion* can be examined. The rocks are patchily dolomitised and slightly cavernous. The axial trace of the syncline has been crossed because the beds now dip at 5° to the south and, furthermore, the syncline is evidently strongly asymmetrical with a southeasterly dipping axial surface.

Walk north along the shore. At [525687] thin, interbedded, pink, crinoidal limestones, nodular limestones with corals and productids, sandstones and conglomeratic sandstones of D_2 age are exposed.

The Lower Carboniferous succession records the submergence of an area of Precambrian to Ordovician rocks and the rapid establishment of a carbonate shelf. The intercalations of shales, sandstones and granule conglomerates record occasional invasions of terrigenous clastic sediments onto the shelf.

Locality 5: Llanfair-is-gaer Church

Drive to the church where a car can be left in the lay-by at the graveyard [502659].

Walk down to the foreshore and onto a wave-cut pavement in a Tertiary mafic dyke 40 m wide and trending at 320°. This thick dyke has a coarse-grained, gabbroic interior and a chilled marginal facies, and is spheroidally weathered. The country rock on the northeast side is a baked and bleached conglomerate, well exposed in low cliffs a few metres to the northeast of the dyke. It consists of sub-rounded to sub-angular pebbles up to 10 cm long in a sand matrix. The clasts are of rhyolite, porphyritic andesite, sandstone and conglomeratic quartzite. The rock is well bedded and imbrication of the more discoidal pebbles is apparent, which indicates current directions from east to west. Thin marls and sandy marls are interbedded with coarse-grained conglomerates, and sun cracks in the marls are infilled with sand. The matrix of the conglomerates becomes red and purple away from the 40 m dyke. Twelve smaller Tertiary dykes, ail less than 1 m thick and trending northwest, cut the conglomerates in the low cliff to the northeast (figure 31). The sediments are clearly subaerial and, from comparison with similar rocks on Anglesey, are placed in the Red Measures of the Carboniferous.

Locality 6: Twt Hill

Drive to Twt Hill in Caernarfon [482631].

The hill is formed by a granite emplaced in the Arvonian volcanics and is itself Precambrian in age. The abundant exposures show the rock to be

Figure 31. Tertiary basic dykes cutting Upper Carboniferous conglomerates in cliff near Llanfair-is-gaer church.

very variable. The main rock type is a coarse-grained, pink facies consisting of pink potassium feldspars up to 10 mm long, smaller cream-coloured plagioclase, and opalescent quartz. This appears to grade over distances of just a few metres into a porphyritic, microcrystalline facies with strongly resorbed quartz phenocrysts up to 5 mm in diameter. The granite is cut by many shears.

Locality 7: The Afon Seiont Section

Park in the castle car park at Caernarfon.

Examine the Arenig beds exposed beneath the castle wall. They consist of very thin, fine-grained sandstones, siltstones and shales dipping 23° northwest. Cleavage (S_1) dips at 55° in the same direction and the beds presumably face upwards. Graptolites indicate the zone of *D. extensus.*

Cross the bridge to the western bank and walk along the road towards the headland. Exposures at the roadside consist of cleaved, muddy sandstones thought by Elles (1904) to be pre-Arenig and to be separated from the castle exposures by the Bangor–Caernarfon Fault. Bedding can be made out and the exposure contains a syncline plunging at about 25° to the northeast. No fossils have been obtained from the exposure which may well be of Lower Cambrian rocks.

Take the road southeast along the western bank of the Afon Seiont. In a largely overgrown exposure behind a rotting seat by the sea-scouts' building are exposed a series of thinly interbedded, calcareous sandstones, siltstones and mudstones. No fossils have been obtained from these exposures in which the beds dip northwest, as do the beds of similar lithology beneath the castle.

Take to the water's edge. Just beyond the boatyard the beds dip at 37° southeast and a minor anticlinal axial trace has been crossed. The lowest beds are again calcareous sandstones, siltstones and shales which, 20 m further along the shore, become horizontal and much more shaley. Graptolites indicating the zone of *D. extensus* are abundant. In a further 10 m the beds again dip gently southeast, and continue to yield graptolites, including *D. extensus* itself, as far as the bend in the river.

At the bend the lithology remains similar and, apart from slight flexuring, dips remain essentially gently southeastward, but the graptolites are now diagnostic of the zone of *D. hirundo.* Exposures being to deteriorate at the prominent rotting hulk and the river now flows more or less along the strike. Rocks can still be examined at the water's edge.

About 150 m before the gasworks, the river swings across the strike once more. There is then an exposure gap for the next 200 m, before a series of unfossiliferous, hard, calcareous, fine-grained sandstones and siltstones crop out. These are succeeded, opposite the gasworks, by siltstones containing *D. hirundo* (Salter) itself. The section now becomes very difficult to work and it is only too easy to fall into the river!

About 200 m southeast of the gas holder, the siltstones are succeeded by southeasterly dipping, sandy shales with fairly common graptolites of the zone of *D. bifidus,* including the index fossil. Cyclopygid trilobites can be found in the siltier laminae. Similar rocks continue to 100 m before the house near the bridge over the Afon Seiont.

The easiest way off the section is to scramble up through the wood and onto the road back to Caernarfon.

EXCURSION 12. THE ARVONIAN AND CAMBRIAN OF THE VALE OF NANTLLE

The aim of this excursion is to examine the sequence from the Arvonian Volcanic Series into the Cambrian, to study the end-Silurian to Devonian F_1 structures typically developed within the Arfon Anticline, and to examine the lithologies present. The excursion begins near the axial trace of the major structure (possibly just in the northwestern limb) and then progresses through the southeastern limb, terminating within the slates of the Cambrian Slate Belt. The sequence to be seen is given in the Cambrian correlation chart in figure 4 (column 1).

There is perhaps a good half-day's work in the last two localities alone, so that transport is essential if the ground is to be covered in one day.
Use the excursion geological map, figure 32.
Reference: Morris and Fearnsides (1926).
1 : 50 000 OS sheets Nos 123 and 115; or 1 : 25 000 sheets Nos. SH45 and 55.

Locality 1: Clogwyn Melyn [4853]

The locality gives its name to the Clogwyn Volcanic Group (≡ Arvonian Series). A car can be left conveniently at the telephone box [485539]. Take the track and then the footpath to [488537].

Many exposures of rhyolitic, crystal-vitric tuff can be examined in the fields of rough pasture. The rocks are cleaved but a eutaxitic structure can be seen, indicating that the tuffs are ignimbrites. This structure is difficult for the inexperienced to pick out because fiamme are absent. The ignimbrites are massive and unbedded, but the eutaxitic structure indicates that here their dip varies from horizontal to up to 30° northwest. Cleavage is constant and dips at 80° towards 315°.

10 m beyond the gate at [488538], an anticlinal hinge can be distinguished in eutaxitic, rhyolitic tuff. A poor development of columnar jointing can be seen. These features are more obvious near the ruined cottage (distinguished by two small, adjacent fir trees and holly trees). Here, columnar-jointed, rhyolitic, eutaxitic ignimbrites with fiamme indicate the same fold hinge.

At the summit of Clogwyn Melyn the ignimbrites can be seen to be overlain by massive, crystal-vitric tuff in which no eutaxitic structure can be made out. The rocks may be sillars. Some 30 m west of the summit they are succeeded by bedded crystal-vitric and vitric tuffs in which the bedding dips 67° towards 318°, so indicating the northwestern limb of the anticline recognised at the ruined cottage. The cleavage dips 80° towards 320°, suggesting that the anticline has virtually no plunge at this point.

Leave the summit by the footpath which leads back towards the road. In the rough pasture to the southwest, just before the cottages, bedded rhyolitic and rare andesitic tuffs may be seen dipping 40° towards 140°, and therefore a synclinal axial trace has been crossed. The Clogwyn Volcanics are therefore folded on axes trending at about 050°.

Locality 2: Mynydd Cilgwyn [4954]

A car can be left near the telephone box at Carmel. Pass through the gate by the telephone box onto the hillside and follow the track leading south-southwest. This leads to two small quarries, cut in Clogwyn Volcanics, which are rapidly being infilled with refuse.

The rocks exposed, at the time of writing, are strongly weathered, rhyolitic tuffs and agglomerates which appear strongly cleaved. Bedding may be seen at the southeastern end of the lower pit, where a portion of a fold hinge is preserved. The northwestern limb is clearly seen and the maximum dip is 30° northwest. The dip lessens to horizontal at the hinge, but the southeastern limb, unfortunately, is not exposed.

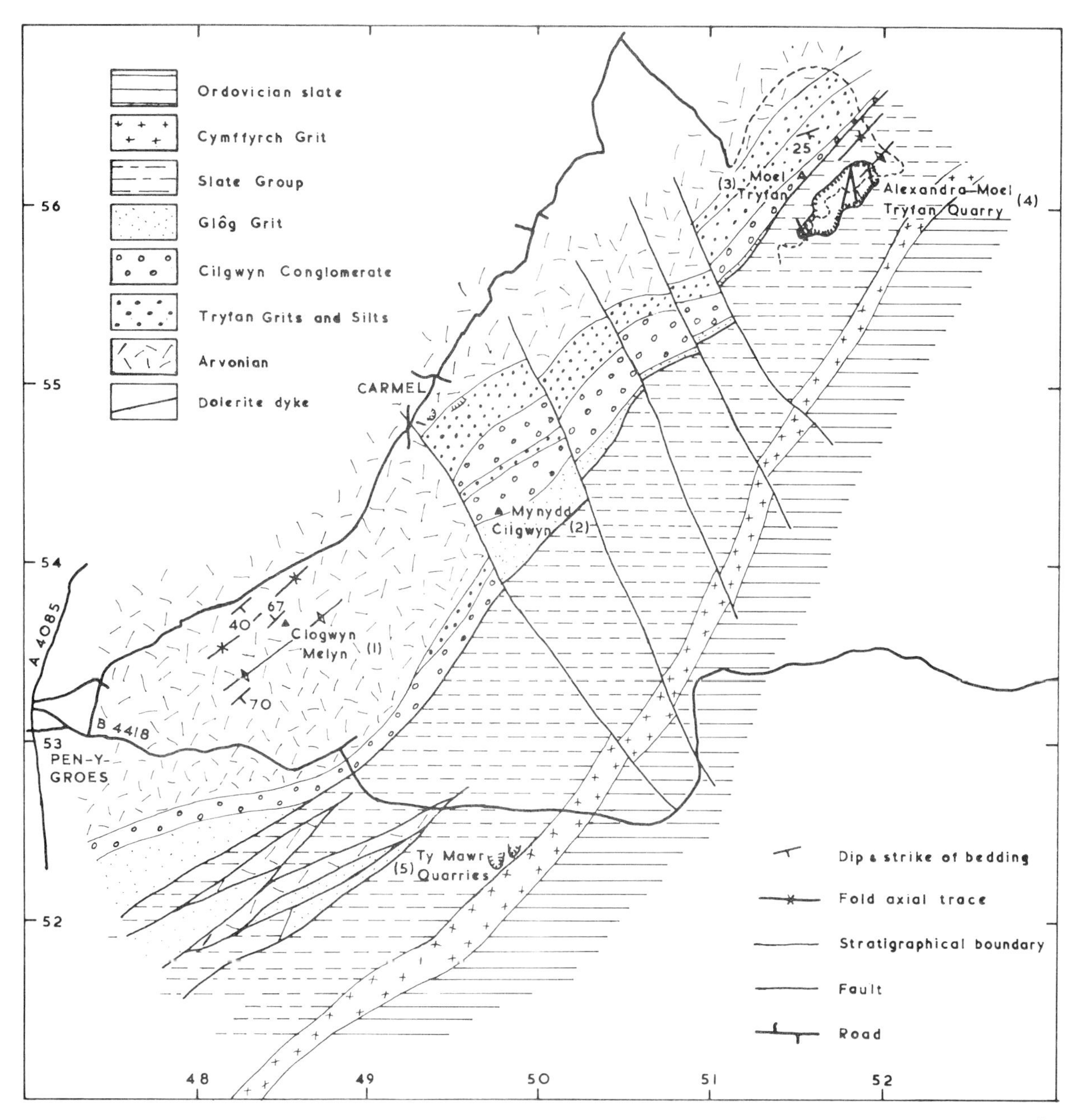

Figure 32. Excursion 12. After Morris and Fearnsides (1926), with permission.

50–70 m along the footpath which runs northeastwards from the top of the quarries, are some dark-weathering exposures of Tryfan Grits. The grits are very strongly cleaved and sericite-rich but the exposures, although essentially *in situ*, have been rotated by hill creep and so no reliable measurements can be made on them. The rocks are conglomeratic sandstones and granule conglomerates, and are very rich in clasts of red jasper derived from the Mona Complex and quartz–porphyry from the Arvonian. Much of the sericite has been derived from detrital feldspars.

Walk up towards the summit of Mynydd Cilgwyn. The Tryfan Silts, which succeed the Tryfan Grits, are not exposed, but sporadic exposures of the Cilgwyn Conglomerate are met with. The best exposures are just northwest of the summit where the rocks can be seen to consist of flattened clasts of rhyolite, jasper and rarer quartzite, in a siliceous matrix. Long axes of clasts, which may be up to 15 cm, lie within the plane of cleavage. Cleavage is strong and dips vary from vertical to 80° northwest. Bedding has not been distinguished.

Immediately southeast of the summit many exposures of the Glôg Grit can be examined. The junction with the underlying Cilgwyn Conglomerate is seen to be transitional. The rocks are mainly greyish white and greenish white, feldspar–quartz sandstones, often coarse and microconglomeratic. They are very strongly cleaved and some specimens have been so strongly deformed and recrystallised that they are now in the form of quartz–chlorite–sericite schists and phyllites.

Locality 3: Moel Tryfan [5156]

Drive to the end of the metalled road and there is ample parking space at [511562]. Climb up the hill, keeping to the north of the slate tips.

Examine the rare, isolated exposures, more or less *in situ*, of laminated, fine-grained, vitric tuffs on the hillside. These rocks weather with a white, silica patina, and are tuffs and tuffaceous sediments within the Tryfan Silt Group. At the summit, examine the superb exposures of Cilgwyn Conglomerate dipping to the southeast. At the base of the crag are finer-grained conglomerates which rest on Tryfan Silts. The siltstones and fine-grained sandstones are cross bedded and show soft-sediment deformation in the form of slump structures. **Exposures are limited and should not be hammered.**

Proceed for 320 m on bearing 005°. Several outcrops of the Tryfan Silt Group show obvious bedding dipping 25° towards 160°. They are medium- and fine-grained sandstones with silty laminae. Some sandstones are obviously tuffaceous with clasts of rhyolite and grains of feldspar and are cross bedded.

Outcrops of the Tryfan Grits further downslope have less fine-grained material and clasts of jasper are common. Coarse, tabular cross bedding can be seen in these coarse-grained, feldspathic sandstones.

The evidence indicates that the Tryfan Grits and Silts, Cilgwyn Conglomerate and the Glôg Grit were laid down in shallow water, often under the influence of strong currents. A minor volcanic episode is recorded by the tuffaceous nature of the Tryfan Silts. Most of the clastic components were derived from late Mona Complex and Arvonian rocks.

Locality 4: Alexandra–Moel Tryfan Slate Quarry [5156]

Descend to the track fringing the northern slope of Moel Tryfan and follow it into the Alexandra–Moel Tryfan slate quarry.

As the quarry is entered, slate waste is seen overlying glacial sands, gravels and boulder-bearing gravels. On the southeastern side of the quarry, boulders are held in a crudely stratified matrix of sand and gravel which contains fragments of marine shells; on the northern side, in a section 5 m thick, a lens of clean sand within the coarser drift contains many recognisable fragments of bivalves, and far-travelled pebbles including coal, chalk and flint. The exposure of Irish Sea Drift is historically interesting, and has been described by Trimmer (1831).

The track enters the quarry proper by way of a cutting through a crag strengthened by a dolerite dyke. The dyke trends at 355° and is vertical. Continue along the track and then look back at this northeastern end of the quarry (figure 33). The quarry is seen to be cut into an anticlinal structure, complicated by strike faults. The deep pit with the pump and ladders is cut in the anticlinal hinge in the Purple Slate. The greywackes of the Dorothea Grit form the northwestern wall of the quarry and can be seen dipping steeply northwest; on the southeast at the track, they dip southeast and have been thinned tectonically. A thin, boudinaged

Figure 33. Photograph showing the essentially anticlinal structure at the northwestern end of the Alexandra–Moel Tryfan quarry.

greywacke forms an upstanding pillar, southeast of which come green slates. Above the thinned Dorothea Grit on the southeastern limb come the southeasterly dipping striped Blue Slates.

Continue along the track to where, near the lowest point, fallen blocks of Dorothea Grit show graded bedding, cross lamination, convolute lamination and a number of sole structures including flute casts, groove casts and load casts. The coarser conglomeratic beds often contain mud pellets. These are turbidite sandstones.

Continue up the track to the southeastern side of the quarry. The dip in the Purple Slate Group is to the southeast and the group is overlain by the greenish Dorothea Grit and the Striped Blue Slate Group. Thin (5 cm thick) grits within the Purple Slate Group develop prominent load casts and are commonly graded. A wealth of sedimentary structures may be observed here and, where they are involved in minor folds, the grits show extreme attenuation in the fold limbs. Green reduction spots in the Purple Slate Group indicate the extent of deformation which the rocks have undergone. The semi-axes can be readily measured and a realistic result can be obtained because we may assume there was no ductility contrast between the enclosing slate and the green reduction spots which, we may further assume, were originally spherical.

Cross back to the northwestern side of the quarry and examine the prominent anticline in Dorothea Grit (figure 34). Note particularly the joints and quartz-filled tension gashes, the strong axial-planar fabric in the greywackes, the thinning of the greywacke beds in the limb, the fault on the southeastern limb that terminates the fold against northwesterly dipping purple slates and the way in which hill creep has modified the attitude of the slates next to the ground surface.

Further up the track, near the southeastern entrance, a dolerite dyke crosses the path. The Purple Slate Group hereabouts contains a large number of 15–20 cm thick, graded greywacke beds which cause obvious refraction of the cleavage.

The slates, greywacke siltstones and greywacke sandstones exposed in the Alexandra–Moel Tryfan quarry record a deepening of the sea following the accumulation of the Glôg Grit. The sedimentary structures of the siltstones and sandstones indicate deposition from turbidity currents.

Figure 34. An anticline in the Dorothea Grit exposed on the northwestern side of the Alexandra–Moel Tryfan quarry (see text for explanation).

Locality 5: Ty Mawr Quarries [497523]

Drive via Pen-y-groes to [496526]. Take the track south, and then the footpath southeast up to the Ty Mawr quarries.

In the Ty Mawr quarry, the back wall is formed by the Cymffyrch Grit (≡Bronllwyd Grit) and a passage, partly obscured by debris, has been demonstrated by Cattermole and Jones (1970) from green slates via grey slates, then siltstones with fine-grained sandstones, and finally into grey sandstones of the Cymffyrch Grit. The passage beds are about 6 m thick. The greywacke units rapidly thicken and coarsen and become conglomeratic in places, with clasts of mudstone, quartz, jasper and feldspars. The greywacke beds are commonly about 50 cm thick and most consist of a graded interval followed by an interval of parallel lamination. They probably represent proximal turbidites.

EXCURSION 13. Y GARN TO MYNYDD TAL-Y-MIGNEDD

The main aim of this excursion is to study the tectonics of this strongly deformed northwestern limb of the Snowdonia Syncline and to examine the Upper Cambrian lithologies. The excursion also provides a magnificent day's hill walking.

The sequence is tabulated in the Cambrian correlation chart in figure 4 (column 1) and the Ordovician correlation chart in figure 5 (column 5).

A car can be left for the day in the lay-by on the B4418 at [537534] near the turn-off to Tal-y-mignedd.

Use the IGS 1:25 000 Geological Special Sheet (Central Snowdonia) as an excursion map.

Reference: Shackleton (1959).

1:50 000 OS sheet No. 115; or 1:25 000 sheet No. SH55.

Part 1: Drws-y-coed to Y Garn Summit

The transport having been parked in the lay-by, look northeast to the cliffs and screes of Craig-y-bera formed by the riebeckite-microgranite intrusion of Mynydd Mawr; and look east to Clogwyn-y-garreg where three massive quartzites of Ffestiniog age form pronounced features and can be seen dipping northward in the northwestern limb of the Drws-y-coed Pericline.

The route to be taken lies to the southeast, up Clogwyn-y-Barcut and the shoulder of Y Garn. Cross southeast over the flat ground to the crags of heavily quartz-veined Ffestiniog quartzite. The quartzite is conglomeratic and massively bedded and here, with dips of 40° to the north, lies in the northern limb of the Drwys-y-coed Pericline.

Climb up over the quartzite towards Clogwyn-y-Barcut, and it is apparent that the quartzites are underlain by tough, splintery, greenish grey slates, siltstones and fine-grained sandstones. Continue to a very prominent crag above a fence where both bedding and the slaty cleavage (S_1) are deformed by a set of brittle, open folds with southeasterly dipping axial surfaces. These are F_2 structures and are only very sporadically developed. Continue around the corner of the exposure where minor F_1 folds are developed in sandstones, siltstones and silty slates. The axial-planar cleavage, S_1, is a very close fracture cleavage in the sandstones and a very strong crenulation cleavage in the finely laminated, muddy siltstones.

The conglomeratic quartzites sometimes show large-scale cross bedding, whereas the thin, fine-grained, well sorted sandstones and siltstones may show ripple-drift bedding, parallel lamination and convolute lamination. The sequence was deposited under the influence of strong currents, well within the reach of wave base, and the presence of the conglomeratic quartzites towards the top of the succession suggests a further shallowing with the possible establishment of sub-littoral conditions.

From the summit of Clogwyn-y-Barcut the three quartzites of Clogwyn-y-garreg show up well. Looking towards Y Garn, the foreground as far as the wall is occupied by Ffestiniog slates, siltstones and sandstones, which are largely grass-covered. Above the wall rise grey crags of Ffestiniog quartzite forming the southeastern limb of the Drws-y-coed Pericline and therefore the axial trace lies in front of us passing through the grass-covered ground. Above the quartzite crags come black outcrops of Llanvirn slates which are faulted against the quartzites, and finally the summit is capped by grey, massively jointed tonalite.

Cross the largely grass-covered ground to the wall at the foot of the spur up to Y Garn. The Ffestiniog quartzites above the wall are strongly cleaved, but here and there contain thin beds of siltstone and fine-grained sandstones which show that bedding dips 62° towards 156°. Cleavage is now vertical. Clearly the axial trace of the Drwys-y-coed Pericline has now been crossed and the exposure lies in the southeastern limb of the fold.

Climb up the spur over the quartzites. They are overlain by a few metres of tough, splintery, grey-green slates and siltstones which are devoid of fossils, but are of presumed Ffestiniog age.

From this point on the spur look southwest across to Trum y Ddysgl and see the three Ffestiniog quartzite horizons curving up the western wall of the cwm. The beds lie in the southeastern limb of a second periclinal fold called the Cwm y Ffynnon Pericline by Shackleton (1959) and, towards the bottom of the exposure, the beds are clearly overturned towards the southeast (figure 35).

The tough, grey-green, silty slates are faulted against the succeeding beds which are soft, blue-black Llanvirn slates. These continue almost to the summit of Y Garn which is capped by a grey-weathering, feldspar-porphyritic microtonalite. The rock is massively columnar-jointed, the attitude of the joints suggesting the intrusion to be a sill-like sheet, dipping east. The junction with the underlying Llanvirn slates is visible near the top of the wall of the cwm and confirms the sheet-like form of the igneous mass.

Part 2: Y Garn to Cwm y Ffynnon

From Y Garn walk 900 m south along the ridge of Clogwyn Marchnad to the summit of Mynydd Drws-y-coed. The summit consists of Llanvirn slates and, at the stile over the fence, minor F_1 folds in the slates plunge 35° towards 055°. The axial-planar S_1 cleavage is here a crenulation cleavage.

Carry on to the col between Mynydd Drws-y-coed and the summit of Trum y Ddysgl. Descend the steep, grassy back wall of the cwm. Examine

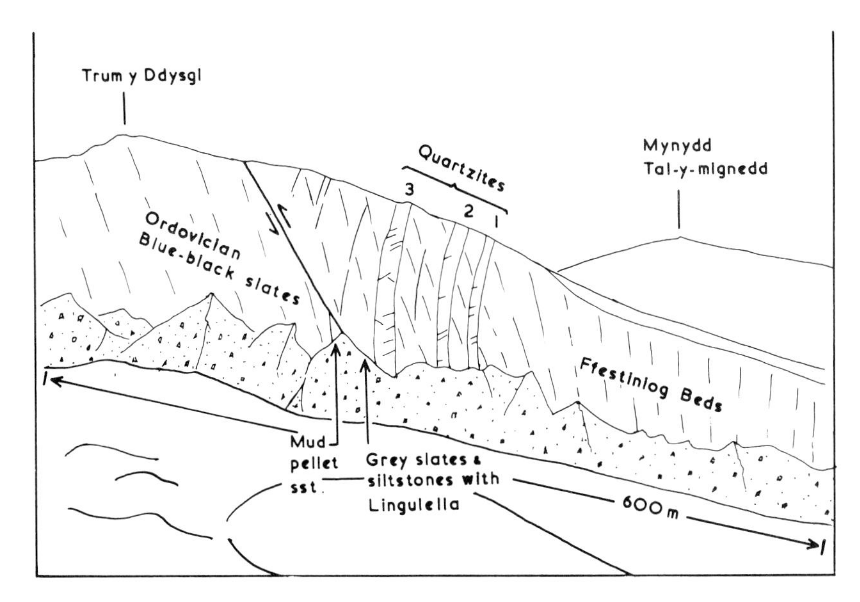

Figure 35. View looking from Y Garn towards Trum y Ddysgl, drawn from a photograph and inspired by Shackleton (1959; figure 2, p 220). The three quartzites are overturned where they crop out just above the scree.

the generally northwesterly dipping Llanvirn slates, with their minor F_1 folds, exposed on the east as you descend. On the west, examine the southeasterly dipping dolerite sheet. This strongly discordant intrusion shows good columnar jointing. A small fault in the back wall of the cwm displaces the dolerite sheet a few metres.

Traverse northwest across the sloping wall of the cwm towards the cliff with the three quartzites. The blue-black, Ordovician slates are underlain by a 2 m thick mud-pellet sandstone which may be Arenig in age. Hereafter, the succession, which is Ffestiniog in age, is as Shackleton (1959) has described it. The sandstone is faulted against 12 m of splintery, grey slates with interbedded siltstones. *Lingulella* is common and the beds may represent the *Lingulella* band. The bedding/cleavage relationship indicates that the beds are overturned to the southeast and this is borne out by cross lamination. The underlying beds comprise 2 m of slates with thin quartzite ribs and then comes the youngest quartzite, (3) in figure 35, which is 20 m thick and conglomeratic. This is underlain by about 25 m of slates with thin quartzites before the second conglomeratic quartzite (2), about 12 m thick, is reached. A further 10 m of slates with thin, interbedded quartzites and sandstones intervene before the oldest quartzite (1) is reached. This last quartzite is about 8 m thick and is underlain by tough, silty, grey slates with thin sandstone ribs. The quartzites are cut by quartz-filled tension gashes and the thinner quartzites within the slates show incipient boudinage.

It is seen therefore, that not only is the southeastern limb of the Cwm y Ffynnon Pericline overturned to the southeast, but also that it is cut by a reverse fault which, in the same sense, carries the Ffestiniog Beds over the Ordovician slates.

Traverse around the northern spur of Trum y Ddysgl and contour around to the head of Cwm y Ffynnon. Here, just to the east of the lowest point of the col between Mynydd Tal-y-mignedd and Trum y Ddysgl, a notch can be seen in the skyline. Shackleton has shown that the notch is cut in Dolgelly Beds (marked as dolerite on the 1:25 000 IGS sheet) which are faulted against Llanvirn slates to the east. The Dolgelly Beds are seen to be slightly overturned where they reach the foot of the cliff where they yield a diagnostic trilobite–brachiopod fauna. The beds are typically black slates and mudstones. Below the Dolgelly Beds come about 30 m of grey-blue, striped, silty slates which contain *Lingulella* towards the top and are assigned to the Ffestiniog Beds. Two boudinaged lenses of quartzite occur in the cliff, below which thin sandstones occur within the silty slates. The hinge of the periclinal fold is reached in about 175 m, beyond which the Ffestiniog slates and sandstones dip at 50–60° west.

Part 3: Cwm y Ffynnon to Tal-y-mignedd

Contour around the west side of Cwm y Ffynnon, across the westerly dipping Ffestiniog slates and sandstones, onto strongly cleaved, acid rock interpreted by Shackleton as part of an intrusive sheet of microtonalite. Cleavage is so intense that the original nature of the felsic rock cannot be determined in the hand specimen.

Continue across poorly exposed ground to the trial levels at [534522] and higher up the hillside at [532519]. At the upper level, excellent specimens of galena, chalcopyrite, arsenopyrite and sphalerite can be picked up. The adits have been driven into a sheet of porphyritic microtonalite. The junction with underlying blue-black Llanvirn slates can be seen immediately north of the adit and the contact can be seen dipping at 40° southeast. The sheet occupies the core of a syncline lying between the Drws-y-coed Pericline on the north, and the Cwm y Ffynnon Pericline on the southeast.

Walk northwest across about 120 m of Llanvirn slates until a fault trending eastnortheast on the southeastern limb of the Drws-y-coed Pericline brings in tough, splintery, grey-black Ffestiniog slates. Continue northwest across the southwestern closure of the pericline to the prominent exposures of Ffestiniog quartzite at [527584]. The quartzite lies in the northwestern limb of the pericline and dips at 60° northwest. Follow the strike of the quartzite northeastward obliquely down the hillside to Tal-y-mignedd and the lay-by where the transport was left.

EXCURSION 14. MOEL HEBOG

This excursion examines the Caradoc succession of sedimentary, volcanic and sub-volcanic rocks in the Moel Hebog Syncline. The sequence is tabulated in the Ordovician correlation chart in figure 5 (column 5).

Use the IGS 1:25 000 Geological Special Sheet (Central Snowdonia) as an excursion map.
Transport can be left for the day in the car park in Beddgelert. The excursion involves about 12 km of very pleasant hill walking. It can be broken up into three parts for the purpose of description.
Reference: Shackleton (1959).
1:25 000 topographic sheet, SH54; or 1:50 000 sheet No. 115.

Part 1: Beddgelert to the Summit of Moel Hebog

Leave Beddgelert on the Caernarfon road (A487). At [585484] take the narrow road leading southwest over the bridge towards Cwm Cloch. Having passed through the small stand of conifers at [581479], leave the road and take the footpath leading northwest. Contour along for about 500 m to the bottom of the gentle spur at [578483] which leads up to Moel Hebog.

Parallel-laminated cleaved sandstones and siltstones of Caradocian age are exposed, which Shackleton has grouped together as representing the horizons elsewhere developed as the Pren-teg Grit, Gorllwyn Slate and Gorllwyn Grit. Bedding dips 62° towards 332°. A strong cleavage, S_1, dips 75° towards 302°. The exposures lie in the southeastern limb of the Moel Hebog Syncline.

Begin the climb up the shoulder. After 400 m notice that the rocks of the same lithology now dip at 45° towards 120° and that a minor synclinal axial trace has therefore been crossed, plunging at about 20° towards the northeast. Carry on up the hill to outcrops 50 m below a point where the track passes through an iron gate in the wall. Bedding in medium-grained, parallel-laminated sandstones dips 40° towards 197°. About 200 m above the wall the sandstones are interbedded with siltstones and occasional white-weathering, rhyolitic, vitric tuffs averaging 5 cm in thickness. These beds are overlain by very coarse-grained, cross bedded sandstones containing rhyolitic clasts up to 3 cm across. The larger clasts are rounded whereas the smaller fragments are still angular; clearly neither are far-travelled. The sandstones are calcareous and show carious weathering.

In a further 100 m the track becomes marked by a prominent series of cairns. The sediments revert to parallel-laminated and rippled sandstones and siltstones. A minor anticlinal axial trace has been crossed because the bedding now dips into the main Moel Hebog Syncline, the sediments dipping 35° towards 277°.

As the prominent cliff formed by the Pitt's Head ignimbrite is neared, the sandstones become coarser-grained and again are interbedded with white-weathering, rhyolitic, vitric tuffs.

The parallel-laminated slates and siltstones of the lower part of the sedimentary sequence accumulated under relatively quiet marine conditions; but strong current action and shallowing is indicated in the upper part, where coarse-grained tuffaceous sediments, often strongly cross bedded, become dominant. Volcanic detritus forms the bulk of the coarser sediments.

Look towards the cliff of Pitt's Head ignimbrite and observe the prominent nodular horizon developed above the base. Nodules are the size of coconuts and cannon balls. Observe also the prominent fault, downthrowing a few metres to the south, which cuts the ignimbrite sheet.

Continue up the spur to the base of the Pitt's Head ignimbrite sheet. The basal facies is seen to be a crudely bedded, strongly cleaved, non-welded, rhyolitic, pumice-lapilli tuff in which the pumice lapilli are flattened in a plane parallel to the cleavage. This passes up into a facies in which the pumice lapilli are flattened in a plane parallel to the base of the sheet, and so a eutaxitic structure is developed. The rock is a feebly welded, columnar-jointed tuff, and is succeeded by the prominent nodular horizon in which nodules reach up to 30 cm across. The eutaxitic structure may be seen to pass through them indicating that the development of the nodules post-dates the welding. The tuff overlying the nodular horizon is intensely welded and has a very pronounced eutaxitic structure with strongly flattened pumice lapilli and possibly fiamme of other origins.

Careful inspection of the exposures indicates that the Pitt's Head ignimbrite here consists of at least two successive ash flows, a lower and an upper, which have cooled separately because the intensely eutaxitic tuff is overlain by a non-welded, rhyolitic, lapilli tuff representing the base of a second flow. This is succeeded by a nodular horizon (1 m thick) in which the relatively small nodules range up to 5 cm and in turn by 3–4 m of eutaxitic welded tuff. (The 1:25 000 map indicates that a thin slate is present within the Pitt's Head ignimbrite at about this horizon southwest of Moel Hebog.)

The Pitt's Head ignimbrite is overlain by an agglomeratic, rhyolitic tuff representing the lowest member of the Lower Rhyolitic Tuff Group. The blocks are of Pitt's Head ignimbrite (including both the strongly eutaxitic and the nodular facies) as well as of rhyolite and sediments. The formation is strongly cleaved and crudely bedded, bedding dipping 35° towards 302°.

On reaching the shoulder, turn southwards up the last rise to the summit of Moel Hebog. Examine the lithologies of the Lower Rhyolitic Tuff on the way. Note the trough cross bedded tuffs and tuffaceous sandstones containing pumice bombs and blocks up to 10 cm across, and the parallel-laminated and sometimes rippled, thin, white-weathering, rhyolitic, vitric tuffs which sometimes show load casts at the base of the beds. Clearly much of the Lower Rhyolitic Tuff was water-lain in this vicinity.

The sequence therefore records an initial shallowing of the marine conditions, with slates giving way to coarse-grained conglomeratic volcanic sands and tuffs. The process culminated with the emergence and emplacement of the Lower Pitt's Head ignimbrite under subaerial conditions. A short-lived subsidence led to the accumulation of a thin conglomerate in the south of the Moel Hebog Syncline. This was followed by re-emergence and the emplacement of the Upper Pitt's Head ignimbrite, again under subaerial conditions. Further subsidence followed because the succeeding tuffs and volcanic sediments of the Lower Rhyolitic Tuff Group were clearly laid down in shallow-water marine conditions and have yielded a Lower Longvillian fauna in the western limb of the syncline.

The summit is formed by a swollen, sill-like intrusion of flow-banded and flowage-folded rhyolite. There are good exposures, including autobrecciated rhyolite, 100 m down dip, west of the summit.

Part 2: Moel Hebog Summit to Bwlch Meillionen

Before starting this next part of the excursion, look northwest down to Bwlch Meillionen and see the hinge of the Moel Hebog Syncline outlined by the thick basalt lavas (figure 36).

Descend northwestward from the summit of Moel Hebog towards Bwlch Meillionen. There is a convenient wall to follow should the cloud be low.

At first the rock is the flow-banded, folded and brecciated, intrusive rhyolite but then, if you are keeping close to the wall, the route takes you back into the underlying, bedded tuffs of the Lower Rhyolitic Tuff Group and then back again through the rhyolite sill. A small thickness of bedded, rhyolitic tuff overlies the intrusion and bedding dips 35° towards 294°, so the traverse still lies in the southeastern limb of the Moel Hebog Syncline.

Figure 36. Photograph of the Moel Hebog Syncline taken from the northwestern slope of Moel Hebog. The trace of the syncline is outlined by the crags of the basalt lava flow.

As the wall curves to the west the succession is ascended and the overlying Middle Basic Group begins here with a magnificent basic agglomerate. The rock is crudely bedded and consists of bombs and blocks of basalt (up to 50 cm across) and lapilli in a matrix of basaltic tuff.

60 m before the wall, which leads southwest down Cwm Llefrith, is an intrusion of strongly vesicular dolerite. Vesicles are often infilled with chlorite, calcite and, less commonly, quartz. The rock looks very much like a vesicular lava, but mapping indicates that it is intrusive. Continue across the vesicular dolerite and so back onto the Middle Basic Group which now consists of tuffs and fine lapilli tuffs of mainly basaltic composition, although thin, rhyolitic, vitric tuffs are also present. Bedding dips 45° towards 297°, so the traverse is still in the southeastern limb of the major syncline.

Cross the wall which runs northeast through Bwlch Meillionen to the first exposures 30 m northwest of the wall. A thin, unmapped, columnar-jointed, dolerite sill is overlain by cleaved, tuffaceous sandstones and siltstones of the Middle Basic Group. Ascend the prominent cleft to the first thick, obvious lava. An agglomeratic tuff is exposed at the base of the crag and is succeeded by a blocky, basaltic lava in which pillows up to 1 m across can be distinguished at the base of the eastern wall of the cleft. Continue through the cleft and the blocky pillow lava is seen to be succeeded by a massive basic, agglomeratic, lapilli tuff and then by a well bedded, basic tuff. These pyroclastic units are followed by a 3 m thick, blocky lava and then by plane-bedded and well cleaved, sandy, basic tuffs. The position is now very close to the axial trace of the major syncline; bedding dips 20° towards 288°, and just before a prominent crag of massively jointed, grey-weathering, plane-bedded rock, the axial trace is reached and beds are horizontal.

The pillow lavas are clearly marine and there is no reason to believe that any part of the sequence is other than marine.

Part 3: Bwlch Meillionen to Moel Lefn

Examine the prominent crag of massively jointed, plane-bedded rocks. It consists of bedded, pumice-lapilli tuffs containing occasional blocks of the underlying Middle Basic Group, rhyolite and both acid and basic pumice. The tuffs rest on an uneven erosion surface, but the relief of the surface has been exaggerated by tectonic deformation. The tuffs represent the Upper Rhyolitic Tuff formation.

Climb up the crag. At the top there is a contact against banded and flowage-folded rhyolite which, a few metres further northwest, shows flow brecciation. Flow-banding dips northwest and mapping shows the mass to be a plug-like, rhyolite intrusion. Continue to the summit of Moel yr Ogof where the rocks are obviously columnar-jointed and consist of a spectacular agglomerate with blocks of rhyolite up to 1 m across in a matrix of rhyolitic tuff. The agglomerate has a sub-circular outcrop, is entirely surrounded by flow-banded and flowage-folded rhyolite, and is interpreted as a vent agglomerate. The vent presumably was a source of rocks higher in the Upper Rhyolitic Tuff Group but which have been eroded away.

Descend northwestward to the bwlch and cross back onto the flow-banded rhyolite of the plug and across the largely grass-covered Middle Basic Group. Keep to the western side of the broad ridge, and a small exposure of unmapped, intricately flow-banded, intrusive rhyolite is met before the main outcrop of the Lower Rhyolitic Tuff is reached. The first members of the Lower Rhyolitic Tuff seen are a series of interbedded, thin vitric tuffs and pumice-lapilli tuffs, both of rhyolitic composition. Bedding now dips 50° towards 092° and therefore the traverse now lies in the northwestern limb of the major syncline.

Continue in the direction of Moel Lefn across a small, flow-banded and flowage-folded, intrusive rhyolite (again unmapped) emplaced in a series of finely laminated, vitric tuffs and crudely bedded and nodular pumice-lapilli tuffs. The tuffs can be seen to occupy the core of a minor anticline plunging at 37° towards 024°.

On the gentle climb up to Moel Lefn, cross bedded, rhyolitic agglomerates and agglomeratic tuffs of the Lower Rhyolitic Tuff and then a 2–3 m thick, poorly columnar-jointed dolerite sill are met. The sill contains autobrecciated horizons and shows a very irregular upper junction with overlying rhyolitic agglomerates which suggests it was a very near surface intrusion. A rather thicker, more typical dolerite sill is crossed followed by more members of the Lower Rhyolitic Tuff and finally, at the summit of Moel Lefn, a sheet of rhyolitic rock, dipping east down the hillside, is reached. The lowest exposures of this rhyolite sheet appear to be 'net-veined' in that a meshwork of resistant rhyolitic rock encloses less resistant angular and sub-angular areas, each several centimetres across. The less resistant areas are more vesicular than the mesh, and the whole represents a weathered, autobrecciated marginal facies. The autobrecciated facies gives way to a superbly flow-banded and flowage-folded interior, and there is a perfect development of columnar jointing. Notice that the lower contact with the underlying Lower Rhyolitic Tuff on the west, the attitude of the columnar jointing and the overall attitude of the flow bands, all indicate that the rhyolite is a sheet dipping eastwards.

Walk northeast down the broad ridge leading to Castell across the rhyolite sill which is itself cut by later, more irregular dolerite sheets which were seen earlier on the gentle climb up to the summit of Moel Lefn.

Descend from Castell through the Forestry Commission plantation to the road and so to Beddgelert.

EXCURSION 15. WAENFAWR TO BEDDGELERT

This excursion is included in this guide primarily for those who are unable to walk the hills for one reason or another. It might also serve to occupy a day of otherwise impossible weather. The traverse takes the user from the Arvonian to the mid-Ordovician; from the southeastern limb of the

Arfon Anticline into the Moel Hebog Syncline. The sequence is tabulated in the Ordovician correlation chart in figure 5 (column 6); and the Cambrian correlation chart in figure 4 (column 1).

It is assumed that some sort of transport is available.
Use the excursion geological map, figure 37, and the IGS 1:25 000 Geological Special Sheet (Central Snowdonia).
References: Cattermole and Jones (1970), Shackleton (1959) and Williams (1927).
1:50 000 OS sheet No. 115; or 1:25 000 sheets Nos SH54 and 55.

Locality 1: Parc [527586]

500 m south of the bridge over the Afon Gwyrfai, on the west side of the road, are old quarry buildings where a car can be left. Take the overgrown track leading southwest for 150 m to the quarry face at [526584].

The quarry is cut in a very fresh, welded, rhyolitic tuff of the Arvonian. The rock is a crystal-vitric tuff containing feldspars and corroded quartz crystals in a welded base of devitrified shards. Fiamme are absent and the welded texture is only obvious in thin section. In some places small blocks and lapilli of pre-existing welded tuff are present. At the northern end of the quarry a dolerite dyke with margins chilled against the ignimbrite can be seen.

Locality 2: Betws Garmon

Drive to the telephone box just before Pont-y-betws [533577] where there is space to leave a car on the roadside verge. Take the road leading southwest towards some old slate quarries.

As the tips are approached, purple Cambrian slates are exposed by the roadside. The quarries are rather overgrown, but the purple and green slates can be examined.

Take the footpath leading northeast across the alluvium to [534572] near the bank of the Afon Gwyrfai. The Cymffyrch Grit (≡Bronllwyd Grit), which succeeds the slates is exposed and bedding dips 75° towards 140°. The rock is a plane-bedded, medium-grained, grey sandstone with thin

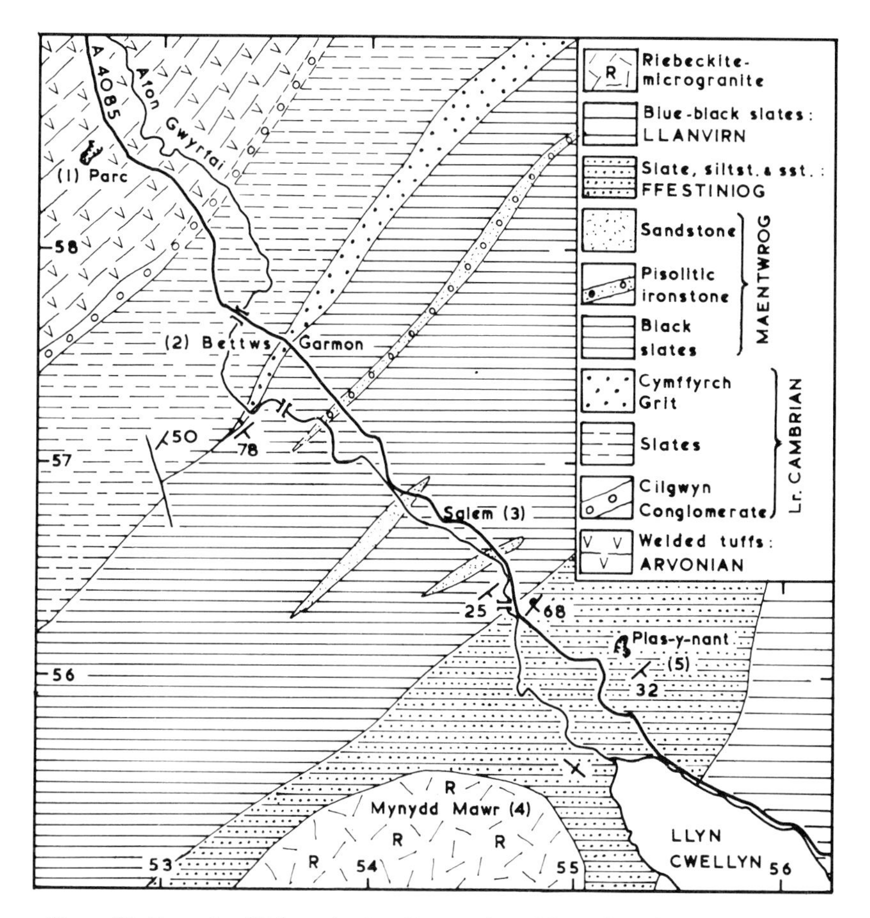

Figure 37. Excursion 15. Largely after Cattermole and Jones (1970), with permission.

siltstone partings. Cattermole and Jones regarded both the lower and upper junctions as faulted.

20–40 m away to the southeast a series of outcrops with small trees growing on them show grey-black, silty slates ascribed to the Maentwrog Beds.

Cross the river by the footbridge and take the footpath to the main road. Turn southeast and in 50 m take the path beside the farm leading northeast to the old mine workings at Ystrad—you can't miss the line of adits sloping up the hillside. The adits are cut into a pisolitic and oolitic ironstone which occurs near the base of the Maentwrog Beds. Pillars of the ore have been left which makes examination easy. Beneath the ore is a worm-burrowed sandstone, which is sometimes conglomeratic. The ore consists of ooliths and pisoliths of a chamosite-like chlorite set in a matrix of siderite, magnetite and chlorite. The ore occurs as a series of sedimentary lenses and bands each up to 5 m thick in a dark, sparsely pisolitic slate and silt. The ore is overlain by uniform, dark, silty slates. The presence of a conglomerate overlain by worm-burrowed sandstone and then by the oolitic ironstone indicates a period of shoaling low in the Maentwrog succession.

Locality 3: Salem

Return to the transport and drive 2 km along the road through Salem and park in a lay-by on the south side of the road at [547563] close to the entrance to Plas-y-nant.

Examine the rocks exposed in the road cutting. These fine-grained, sometimes cross laminated, sandstones and slates have been mapped as representing the Ffestiniog Beds. Just before the driveway to Plas-y-nant they are clearly overturned to the northwest because bedding dips 68° towards 132°, whereas the slaty cleavage (S_1) dips 48° towards 152°. However, at the northern end of the cutting the bedding/cleavage relationships and cross lamination suggest that the beds are the right way up: bedding dips 32° towards 162°, whereas cleavage dips 78° towards 145°. The simplest interpretation of these data is that a syncline with a very strong northwesterly vergence is contained within this roadside exposure (figure 38).

Take the bridge over the river and walk northwest along the old railway track for 40 m, where a fine-grained, pyritous, grey sandstone dipping at 25° to the south is exposed and represents the lowest member of the Ffestiniog Beds.

As the track emerges from the trees, dark slates of the Maentwrog Beds are exposed. In a small quarry 60 m before a house is reached, a tight anticline, with a steep plunge to the southwest, can be shown to be present in silty slates and siltstones.

Figure 38. Roadside exposure of Ffestiniog sandstones and slates. The cleavage (S_1) dips less steeply than the bedding, suggesting that the bedding is overturned.

Locality 4: The Mynydd Mawr Intrusion

Return to the transport and drive to a lay-by on the north side of the road at [553558]. Walk southwest across the flat at the head of Llyn Cwellyn and ford the Afon Gwyrfai. Cross to the crags beneath Castell Cidwm at [550556].

The rocks exposed are well within the contact aureole of the Mynydd Mawr riebeckite-microgranite and are bleached and hardened slates, sandstones and quartzites of the Ffestiniog Beds. The attitude of the bedding has been influenced by the intrusion; it strikes at 128° (almost tangential to the contact) and stands vertically.

Carry on southeast to the Afon Goch. Turn southwest and climb the steep valley to the top of the cataract. Here, the flow-banded and autobrecciated marginal facies of the intrusion can be recognised. The rock is a felsite, weathering grey, but white or pink on a fresh surface and carrying small, dark blue riebeckites.

Locality 5: Plas-y-nant Quarries

Return to the lay-by where the transport was parked. Cross northwestward towards the old Plas-y-nant quarries at [552562].

In and near the quarries are excellent exposures of tough, splintery, grey, grey-blue and grey-green slates with silty laminae and thin sandstone and quartzite beds. These are the Plas-y-nant Beds of Williams which were thought to be possibly Arenig in age. Shackleton suggested they were more likely to be Ffestiniog, and this was confirmed by Crimes (1969a,b) when he recorded the occurrence of *Cruziana semiplicata* (Salter) in beds from this horizon elsewhere in Snowdonia.

Locality 6: Snowdon Ranger [564551]

Return to your transport and drive to the Snowdon Ranger Youth Hostel. Take the Snowdon Ranger track towards Snowdon.

Examine the sporadic exposures of blue-black, soft, micaceous slates along the track. These are the Maesgwm Slates of Llanvirn age which have yielded Llanvirn didymograptids, although only after fairly lengthy search.

Locality 7: Rhyd-ddu [572526]

Drive to the car park south of Rhyd-ddu. Leave the car park at the northern end along the former railway line.

At the gate where the track to Snowdon begins examine exposures of coarse-grained dolerite. Cross to a prominent slot cut in the crags to the southeast. The slot gives access to an old slate quarry at [573526] cut in the Glanrafon Slates of Caradocian age. Beds of silty slate each a few metres thick are separated by thin, irregular, sandy beds. Bedding dips 40° towards 145°; outcrops lie in the northwestern limb of the Moel Hebog–Snowdon Syncline.

A short distance to the southeast the slates are overlain by the Glanrafon Grits. These are coarse-grained, massive sandstones, with large-scale cross bedding. They contain ovoid concretions of silica as well as rhyolite pebbles. The rocks are essentially volcanic sands consisting of abundant andesine and altered rock fragments. They clearly indicate shallowing of the marine conditions.

Locality 8: Pitt's Head [576514]

At the end of the 1500 m straight, southeast of Rhyd-ddu is a low crag projecting from the peat on the west side of the road. This is Pitt's Head: the type locality for the Pitt's Head 'Rhyolite'. The rock is a pumice-lapilli, crystal-vitric tuff of rhyolitic composition which here is strongly welded and possesses a eutaxitic structure due to the presence of fiamme which, the author considers, are mostly pumice lapilli deformed by welding. The rock is very flinty when fresh. A prominent nodular horizon is developed near the middle of the thickness exposed. The eutaxitic structure dips 40° towards 140° and we may take this to be the attitude of the ignimbrite sheet. The ignimbrite was emplaced subaerially and so the uplift which led to shoaling in the underlying sediments culminated with the eruption and emplacement of the Pitt's Head ignimbrite. The tuff is overlain by a dolerite, but then exposures cease against peat and glacial drift.

Locality 9: Pont Cae'r-gors

Park at the Forestry Commission picnic site [576508].

Not only the watershed, but also the axial trace of the Moel Hebog–Snowdon Syncline has been crossed, because in the bed of the Afon Colwyn, south of the car park, and in the fields on the east of the main road, a series of rhyolitic tuffs now dips 48° towards 288°. The rocks are representatives of the Lower Rhyolitic Tuff which overlies the Pitt's Head flows. Bedded, vitric tuffs, lithic tuffs up to 1 m thick, crystal-vitric tuffs, lapilli tuffs and nodular tuffs are all present.

Walk down the road to a small quarry on the east at Llam-y-trwsgl [575504]. The quarry is cut in a splendidly eutaxitic, welded tuff in which the fiamme of deformed pumice lapilli have been chloritised. The rock is shown as the Lower Rhyolitic Tuff on the 1:25 000 geological map, and it may be a welded tuff within the Lower Rhyolitic Tuff, but to the author it has the characteristics of the top of the Pitt's Head ignimbrite as it is usually developed in this area.

Continue for 150–200 m down the road to the Lion Rock where the Pitt's Head flow is strongly welded and the silicified fiamme are very flattened. Dips are about 70° to the west and so the outcrops clearly lie in the southeastern limb of the syncline. The ignimbrite sheet is seen to be underlain in the field to the east by coarse, conglomeratic sandstones which are green in colour and rich in clasts of rhyolite, andesite, pumice and tuffaceous debris. These Glanrafon Grits are in reality water-lain and extensively re-worked tuffs of overall intermediate (andesitic) composition. They were examined earlier in the traverse in the northwestern limb of the syncline near Rhyd-ddu.

A traverse eastward across the rough pasture for about 500 m takes you through, firstly, a dolerite sill, then more sandstones of the Glanrafon Grits, and so onto Glanrafon Slates once more.

Roadside exposures down to Beddgelert remain in westerly dipping Glanrafon slates, sandstones and siltstones on the southeastern limb of the Moel Hebog–Snowdon Syncline.

EXCURSION 16. SNOWDON: CEFN-DU TO CLOGWYN DU'R ARDDU

The aim of this excursion is to complete a traverse from the Arfon Anticline and the Arvonian volcanics, across the northwestern limb of the Snowdonia Syncline, to Snowdon itself. The traverse runs from northwest to southeast, beginning near Llanrug and ending at Llanberis. The excursion affords about 20 km of hill walking and it may well be that for most geologists, two days' work is entailed. The sequence is tabulated in the Cambrian correlation chart in figure 4 (column 2) and the Ordovician correlation chart in figure 5 (column 6).

Use the IGS 1:25 000 Geological Special Sheet (Central Snowdonia) as an excursion map, together with excursion geological map, figure 39.
Reference: Williams (1927).
1:50 000 OS sheet No. 115; or 1:25 000 sheets Nos SH55 and 56.

Part 1: Cefn-du to Llanberis

Access to Cefn-du can be gained by one or other of several footpaths leading southeast from the road from Llanrug to Waenfawr.

Cross the largely unexposed rough ground to the crags at Gareg Lefain [542613] where massive, pink-weathering, rhyolitic, welded tuffs belonging to the Arvonian are exposed. They are readily recognisable because fiamme impart a pronounced eutaxitic structure to the rock and in thin section shards can be seen to be strongly deformed, so defining a good eutaxitic texture. The tuffs are folded on a mesoscopic scale, the geometry suggesting that the exposures lie in the northwestern limb of a major anticline. Folds plunge at 10° towards 070°.

Walk southeastwards up the shoulder towards the summit of Cefn-du across sporadic exposures of eutaxitic, rhyolitic, welded tuff. About 200 m after a fence has been crossed, bedded rhyolitic tuffs crop out, dipping 70° towards 342°. A very fresh, 30 cm thick dolerite dyke cuts the tuffs. It is uncleaved and has a chilled, fine-grained margin and well developed cooling joints. It is clearly post-tectonic, and almost certainly Tertiary.

In a further 50 m strongly cleaved but massive agglomeratic lapilli tuffs crop out. The rock consists of lapilli of rhyolite, up to 5 cm across, in a matrix of pink-weathering, rhyolitic tuff. The rock continues to crop out until about 40 m northwest of the summit, but the summit consists of

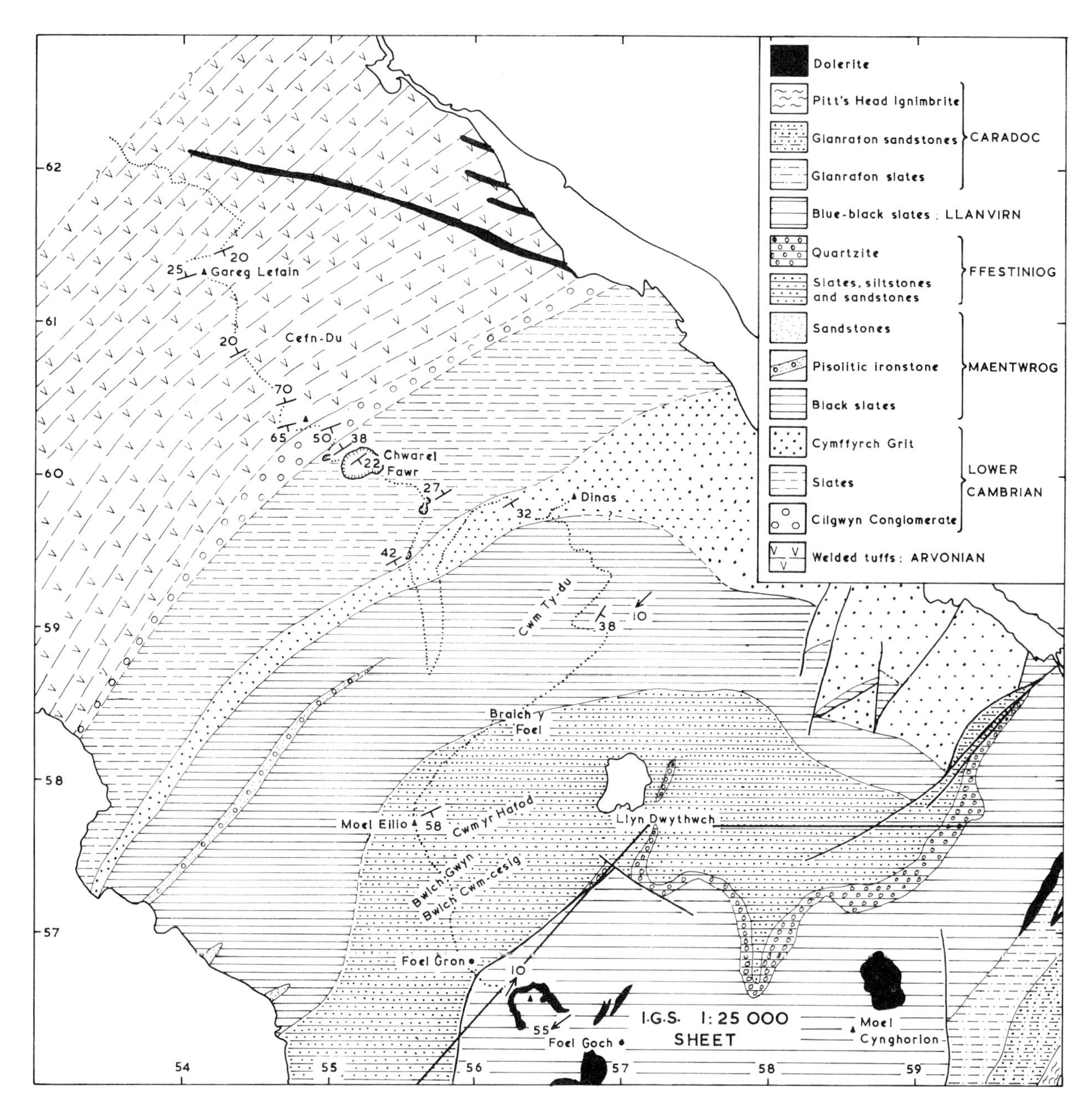

Figure 39. Excursion 16. Partly after Williams (1927), with permission.

rhyolitic, eutaxitic welded tuff in which the eutaxitic structure dips 65° towards 167°, suggesting that the axial trace of the major structure lies to the northwest of the summit.

50 m southeast of the summit are outcrops of a strongly cleaved conglomerate containing deformed pebbles of rhyolitic volcanics, quartz, jasper, and sand-grade clasts of the same materials, together with feldspars and a matrix which is now an aligned, lepidoblastic felt of sericite and chlorite. The formation probably represents the Cilgwyn Conglomerate. In a further 50 m to the southeast the grain size has decreased and the proportion of sericitic matrix has increased. 30 m further on, the rock has passed upward into a silica-cemented, coarse-grained sandstone with rare conglomeratic horizons, dipping at 50° to the southeast. There thus appears to be passage from a Lower Cambrian conglomerate, perhaps the Cilgwyn Conglomerate, into sandstones (perhaps the Glôg Grit).

Continue to the very small slate quarry at [549602] where purple, silty, Lower Cambrian slates are cut by an amygdaloidal dolerite dyke trending at 032°. The amygdales are of chlorite. The dyke is strongly cleaved and was obviously emplaced prior to the F_1 folding. A minor synclinal axial trace runs between this quarry and the previous exposure of Glôg Grit because the slates dip 38° to the northwest.

Cross eastward 100 m to the large, abandoned slate quarry, Chwarel Fawr [552601]. Look across to the eastern face where two 1 m wide dolerite dykes trend at 092° and cut the southeasterly dipping purple slates. On the south of the dykes a fault throws green slates and greywackes against the purple slates (figure 40). A minor anticlinal axial trace must run between this and the previous quarry.

Continue southeast to the quarry on Bwlch-y-groes at [557597]. Purple slates are cut by a readily accessible dolerite dyke varying from 3–5 m in thickness and trending at 092°. The margins are cleaved but the interiors are apparently unaffected. The purple slates dip northwest, and therefore yet another anticlinal axial trace has been crossed between this and the previous quarry.

Walk southsouthwest for 350 m to the old quarry on the hill slope at [556595] where blue-grey and purple-grey slates are exposed. The slates dip southeast and are close to the top of the slate succession.

About 30 m further up the hillside there are many small exposures of the Bronllwyd Grit. Bedding is obvious, but the anomalous dip of the cleavage in many outcrops indicates that most of the small exposures have been rotated by hill creep. It is nevertheless clear that the beds are steeply dipping to the southeast. The Bronllwyd Grit forms a pronounced step, above which is a peat-covered flat with no exposures. There is then a second step up to the signal mast and, 75 m northeast of the beacon and at the mast itself, Maentwrog Beds crop out. The beds consist of strongly cleaved, grey-black slates, siltstones and fine-grained sandstones dipping steeply southeast.

Return northeast down the spur to the track at [561598] where the Bronllwyd Grit is exposed dipping at 32° southeast. The turbidite character of the greywackes is very well demonstrated here. A typical 2 m thick unit consists of a lower, massive, graded interval succeeded by a parallel-laminated interval, and then by a cross laminated interval. Sometimes only the parallel-laminated interval is present. The pelitic interval is often missing and the parallel-laminated interval, which may be convoluted, is usually overlain by the graded base of the succeeding unit.

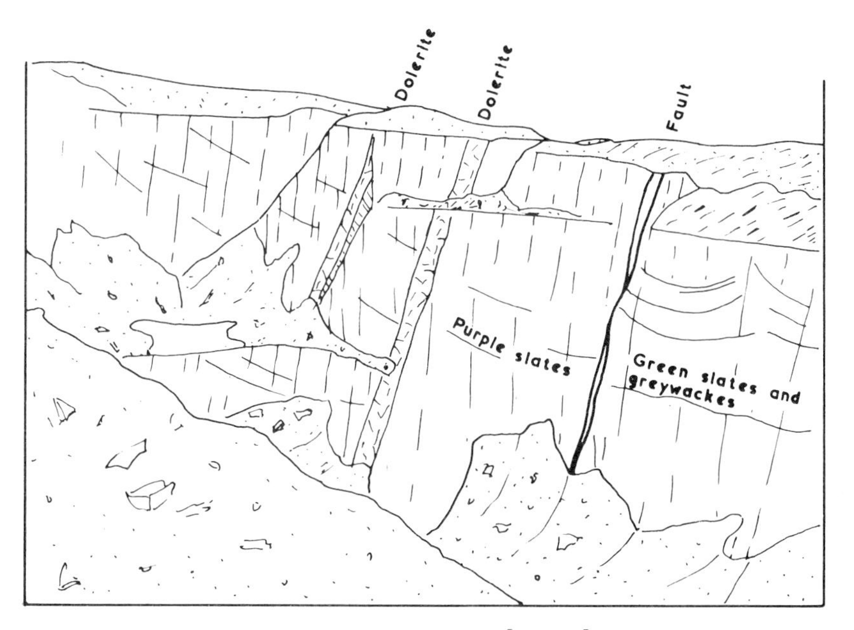

Figure 40. Eastern face of abandoned slate quarry at [552601]. See text for explanation.

Continue down to Dinas at [567598]. (The hut circles southwest of the hill are well worth an inspection.) Bronllwyd Grit forms the hill, the beds consisting of a succession of graded intervals each 50–100 cm thick, with only rarely a parallel-laminated interval. The beds represent proximal turbidites.

Walk south to the exposures at about [568591] on the eastern side of Cwm Ty-du, which consist of very strongly cleaved Maentwrog silty slates, siltstones and fine-grained sandstones, dipping generally at about 40° southeast. Where the track leads off south to Llyn Dwythwch at [572592] the outcrops are of white-weathering, parallel-laminated, fine-grained sandstones ascribed by Williams to the Maentwrog Beds. F_1 folds plunge at 10° towards 228°. The axial-planar cleavage (S_1) dips 65° towards 132°, but a later fracture cleavage (probably S_3) is developed, dipping 80° towards 180°.

If the excursion is broken into two, this is a convenient place to divide it and to descend to Llanberis.

Part 2: Llanberis to Clogwyn du'r Arddu

Climb up onto the end of the spur Braich y Foel [570590].

Williams has drawn the boundary between the Maentwrog and Ffestiniog Beds as cutting slightly obliquely across this northeasterly projecting spur, so that the first siltstones and fine-grained sandstones seen have been mapped as Maentwrog Beds. As the summit of Moel Eilio is neared, so the traverse reaches Ffestiniog Beds which, at the summit can be examined on the eastern face. The rocks are grey, fine-grained sandstones, siltstones and very thin slates, which give rise to prominent pink and grey screes. The beds dip southeast at 58° into the Snowdonia Syncline.

Walk around the top of Cwm yr Hafod to Bwlch Gwyn [559574], where a ridge projects eastwards. The exposures consist of dark grey and greenish grey slates, silty slates and siltstones tentatively thought by Williams to be Arenig, but since shown by Shackleton (1959) and Crimes (1969) to belong to the Ffestiniog Beds. Continue the descent to Bwlch Cwm-cesig and start the ascent to the first and higher of the two peaks of Foel Gron [560568]. The rocks become finer-grained with fewer siltstones and more slates. They are cut by a prominent set of joints which gives rise to pronounced grass-covered ledges on the cliff dipping 30° towards 286°. The slaty (S_1) cleavage is also deformed by a set of open, sideways-facing folds, probably F_2, the axial surfaces of which have the same attitude as the prominent joints.

At the summit of Foel Gron a steep, narrow gully trends at 052° and is probably the site of a fault which, to the south, throws down soft, blue-black Llanvirn slates with poorly preserved graptolites, against the tougher, greenish grey Ffestiniog slates. The wall of the cwm is here very steep and dangerous. 30 m beyond some old iron fence posts, the Llanvirn slates and thin siltstones are strongly folded, the folds plunging at 10° towards 030°. An axial-planar slaty cleavage, S_1, dips at 85° to the southeast. The slates are cut by a strongly cleaved dolerite which was clearly emplaced prior to the F_1 movements and the generation of S_1. The remainder of the excursion lies in ground covered by the IGS 1:25 000 map. On the descent to the col between Foel Gron and Foel Goch an exposure of columnar-jointed dolerite is crossed. 20 m before the lowest point, the Llanvirn slates carry prominent silty laminae and are strongly folded on a mesoscopic scale with folds plunging at 55° towards 237°. The axial-planar cleavage, S_1, in this lithology is a very closely spaced fracture cleavage.

For the next 2·5 km the route over Foel Goch and Moel Cynghorion lies in Llanvirn slates. The rocks are folded on the mesoscopic scale but the absence of any marker horizons makes it difficult to distinguish larger-scale structures. The slates often split readily along the bedding and yield poorly preserved graptolites sporadically.

On the descent from Moel Cynghorion down Clogwyn Llechwedd-llo to Bwlch Cwm Brwynog [591557], look towards Clogwyn du'r Arddu and Snowdon.

The bwlch is occupied by a thin, fault-bounded sliver of Glanrafon (Caradoc) Slates and the ridge up to Clogwyn du'r Arddu consists of southeasterly dipping Glanrafon sandstones overlain by the Pitt's Head ignimbrite, a thin sliver of slate, and then various members of the Lower Rhyolitic Tuff. A minor syncline is developed trending northeast through Clogwyn du'r Arddu and the southeastern limb is faulted. Bedded Pyroclastic Group representatives then occupy the ground to the summit of Snowdon, preserved in the Snowdon Syncline.

At Bwlch Cwm Brwynog the thinly laminated siltstones and slates of the fault-bounded sliver of Glanrafon Slates can be seen to be gently flexed but, overall, the bedding dips 65° towards 237° into the Clogwyn du'r Arddu Syncline. The second fault brings in the Glanrafon Grits which here are interbedded sandstones, siltstones and thin slates. The beds dip uniformly southeast at about 50° and are succeeded by massively bedded sandstones. Towards the top, a tuffaceous sandstone about 6 m thick is rich in lapilli of rhyolite which have been flattened in the plane of the cleavage so that the largest semi-axes range from about 0·5–2 cm. The Glanrafon Grits are rich in volcanic detritus and, in reality, are shallow-water volcanic sands and silts.

The sandstones are followed by the Pitt's Head ignimbrite which is crudely columnar-jointed. This rhyolitic lapilli tuff is intensely welded here with very obvious, strongly flattened fiamme. The ignimbrite was emplaced subaerially and therefore uplift followed the accumulation of the Glanrafon Grits.

Leave the shoulder and descend along the top of the scree at the base of the crags into the cwm holding Llyn du'r Arddu. The Pitt's Head ignimbrite is seen to be overlain by what is essentially an andesitic lapilli tuff but which is indicated on the IGS 1:25 000 geological map as a sandstone. The andesitic lapilli tuff is succeeded by the Lower Rhyolitic Tuff. The formation begins with 4 m of an agglomeratic, rhyolitic, welded tuff in which the fiamme are intensely flattened and deformed around the blocks (up to 10 cm across) of rhyolite. The unit is obviously columnar-jointed. It is followed by a 2 m thick agglomeratic lapilli tuff which is non-welded. Then follows a thick, very coarse-grained, rhyolitic agglomerate in which the blocks range up to 60 cm but are commonly 20–30 cm across. The blocks consist of pumice, rhyolite and ignimbrite and are usually rounded to sub-rounded. The blocks fine upward and the proportion of matrix of rhyolitic pumice increases until the unit passes into a lapilli tuff. It is obviously cleaved.

This unit is succeeded by massively bedded, rhyolitic tuffs with agglomerate horizons about 30 cm thick, rich in blocks and bombs. The tuffs and agglomerates are strongly cleaved.

An horizon follows which is recorded on the 1:25 000 geological map as sandstone but which, in fact, consists of a strongly ripple-marked,

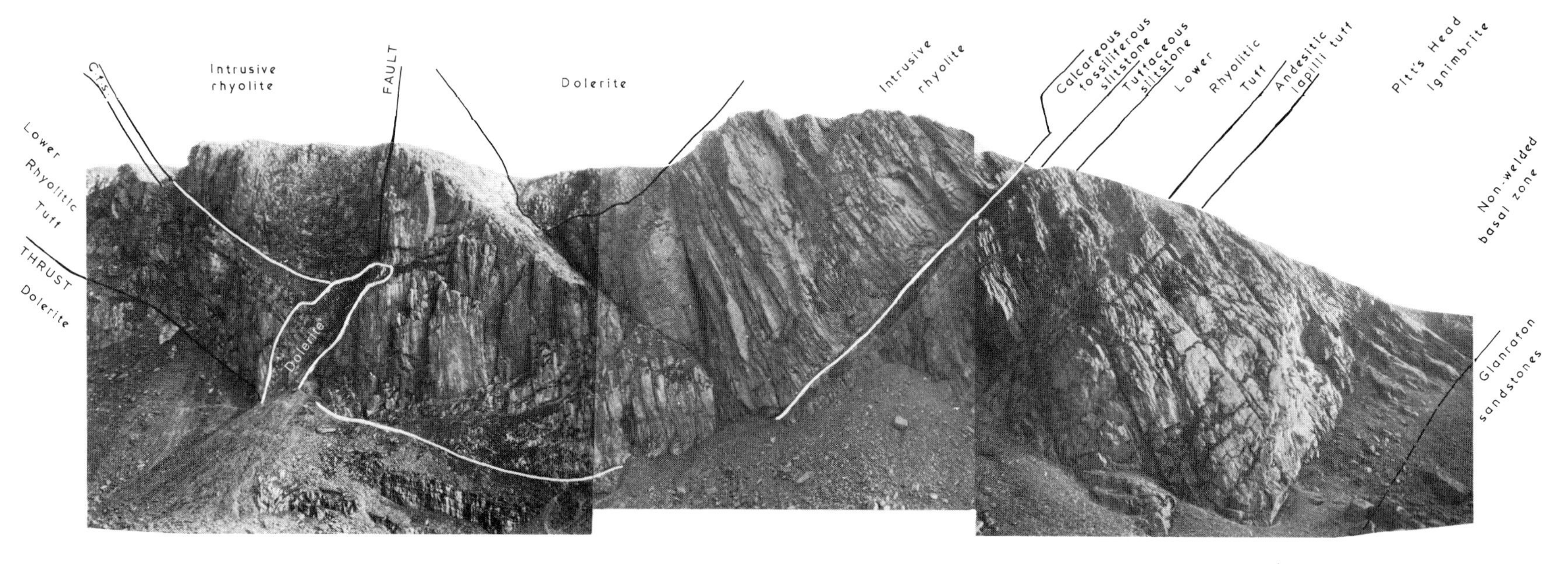

Figure 41. Clogwyn du'r Arddu, showing the succession and the faulted synclinal structure (see text for details).

tuffaceous siltstone which is white-weathering at the base. It is succeeded by a massive, yet laminated, grey sandstone and then by 3–4 m of a blackish blue, calcareous fossiliferous tuff which yields Lower Longvillian brachiopods and trilobites.

The core of the syncline is occupied by a finely banded rock which shows isoclinal, often recumbent folds. This rock is an intrusive rhyolite which has not been separated from the tuffs of the Lower Rhyolitic Tuff on the IGS 1:25 000 map.

Descend to the lake and look back at the cwm wall. The geology is illustrated in figure 41.

The best route back to Llanberis is via the pony track, so the easiest way is to contour around to join the track below Halfway Station.

EXCURSION 17. THE SNOWDON HORSESHOE

A magnificent day's hill walking can be combined with a little geology. The Crib Goch Syncline is regarded by Rast and Bromley as defining the northern arc of the mid-Ordovician rim syncline and is therefore well worth careful examination. The sequence is tabulated in the Ordovician correlation chart in figure 5 (column 6).

Transport can be left for the day in the car park at Pen-y-pass. The route lies up Crib Goch, along the ridge to Crib-y-ddysgl, to Snowdon, Lliwedd, and so back to Pen-y-pass.
Use the IGS 1:25 000 Geological Special Sheet (Central Snowdonia) as an excursion map.
References: Bromley (1969), Rast (1969) and Williams (1927).
1:50 000 OS sheet No. 115; or 1:25 000 sheet No. SH65.

Part 1: Pen-y-pass to Crib Goch summit

The car park at Pen-y-pass is on the Lower Rhyolitic Tuff Group. Leave the car park and walk southwest towards the low summit at the eastern end of the ridge [643552]. Examine plentiful exposures of the Lower Rhyolitic Tuff Group on the way. The rock types are usually poorly bedded, non-welded, pumice lapilli tuffs of rhyolitic composition. Notice that the lapilli are flattened in the plane of the cleavage.

Cross some poorly exposed, largely grass-covered ground which conceals rocks of the Bedded Pyroclastic Group (here faulted against the Lower Rhyolitic Tuff) to reach a conspicuous crag of flow-banded, flowage-folded and autobrecciated rhyolite which, from the mapping, can be shown to be a small intrusion. The summit of the knoll consists of the same intrusive rhyolite with a small cap of bedded, rhyolitic, lapilli tuffs of the Lower Rhyolitic Tuff Group.

Look towards Crib Goch, the summit of which is a plug-like intrusion of rhyolite. Much of the spur leading west up to the summit is in Lower Rhyolitic Tuff, except for a small outlier of grey-weathering Bedded Pyroclastic Group near the bottom of the spur.

Walk along the low ridge towards the Crib Goch spur across interestingly sculpted outcrops of the Bedded Pyroclastic Group. The rocks are well bedded tuffs, sands and silts of basic composition. Their often highly calcareous composition has led to differential erosion of the beds and has given rise to the interesting shapes. Occasional andesitic and rhyolitic tuff horizons prove more resistant. The final knoll on this low ridge consists of rhyolitic lapilli tuff, with the lapilli flattened in the plane of the cleavage, and this is cut by a thin, flow-banded and flowage-folded rhyolite. At the col where the Pyg Track crosses the low ridge, basic tuffs of the Bedded Pyroclastic Group are exposed.

Cross to the start of the spur up to Crib Goch where bedded, rhyolitic tuffs dip 40° towards 116° beneath the Bedded Pyroclastic Group. A minor anticlinal axial trace is then crossed and a small synclinal inlier of Bedded Pyroclastic Group basic tuffs is reached. The traverse then crosses onto the Lower Rhyolitic Tuff once more; first onto bedded agglomeratic and lapilli tuffs and then onto cross laminated, finer-grained tuffs, all of rhyolitic composition. The intrusive junction of the Crib Goch rhyolite is at the point where the spur steepens abruptly. The marginal facies is flow-banded and flowage-folded, and columnar jointing is very well developed. Continue over the intrusive rhyolite to the summit.

Part 2: Crib Goch to Yr Wyddfa

From the summit of Crib Goch look west along the ridge to Crib-y-ddysgl and southwest to Yr Wyddfa (figure 42). The first 400 m of the Crib Goch ridge is over intrusive rhyolite, leading down to the lowest point on the ridge at Bwlch Coch, where the Bedded Pyroclastic Group forms the bwlch and the next 300 m of rising ground. At the second hump on the ridge, on the slope up to Crib-y-ddysgl, a dolerite cuts across the ridge. Beyond this is the largest synclinal inlier of the Upper Rhyolitic Tuff Group in which is emplaced a sill of intrusive rhyolite and several dolerites.

Set out along the ridge across the intrusive rhyolite and the pinnacles which consist of an autobrecciated marginal facies. Notice the prominent, thick quartz veins cutting the rhyolite. At 20 m before the fenced-in patch of experimental pasture, the contact of the intrusion with the Bedded Pyroclastic Group can be examined. For the next 300 m the ridge consists of basic tuffs and lapilli tuffs until a dolerite with excellent columnar

Figure 42. Crib Goch ridge, Crib-y-ddysgl and Snowdon, from Crib Goch summit.

cooling joints is reached. The attitude of the joints indicates that the dolerite sheet dips south towards Glaslyn. Dolerite forms the crest of the ridge for the next 120 m, and this becomes amygdaloidal towards its upper contact, which is against the Upper Rhyolitic Tuff Group. The rocks, which here dip at 30° northwest, are well bedded, massively jointed, crystal-vitric and lithic tuffs with occasional lapilli tuffs.

In a further 50 m there is an abrupt step on the ridge due to the presence of a resistant sill-like intrusion of columnar-jointed, flow-banded and flowage-folded rhyolite. The next 300 m of the ridge are formed from members of the Upper Rhyolitic Tuff and the synclinal axial trace crosses the ridge obliquely here. About 200 m before the summit of Crib-y-ddysgl, the dip of the tuffs is to the east and a thin sill of highly amygdaloidal, bright green dolerite crosses the ridge. The summit consists of another thicker, but similar dolerite. The amygdales are most commonly of chlorite, but calcite and silica, as well as compound, amygdales occur.

Look across from the summit to Snowdon and observe the synclinal nature of the Bedded Pyroclastics forming Yr Wyddfa. From the summit of Crib-y-ddysgl traverse across the southeasterly dipping Upper Rhyolitic Tuff onto the underlying Bedded Pyroclastic Group, and so up to the summit of Yr Wyddfa. An anticlinal axial trace trending at 040° crosses the ridge about 350 m northwest of Yr Wyddfa and the Bedded Pyroclastics turn over to dip at 50° to the southeast. 120 m before the summit station a thin sill of cleaved, amygdaloidal dolerite and an intrusive rhyolite are emplaced in the basic tuffs.

Part 3: Yr Wyddfa to Pen-y-pass

Follow the track from Yr Wyddfa towards Y Lliwedd. The track descends over northwesterly dipping basic tuffs of the Bedded Pyroclastic Group strengthened by two sills of dolerite. At the low point in the ridge, at Bwlch-y-saethau, the basic tuffs are underlain by northwesterly dipping rhyolitic tuffs of the Lower Rhyolitic Tuff Group. Bedding is very difficult to pick up in these strongly cleaved, massive lapilli tuffs. These are the tuffs which Dakyns and Greenly (1905) thought were subaerial and had been formed by Péléean-type eruptions, and which several modern writers interpret as sillars, that is, non-welded ignimbrites.

Continue along the ridge to Bwlch Ciliau and begin the ascent of Y Lliwedd. An anticlinal axial trace is crossed and when bedding can be found it is seen to dip to the southeast. Immediately after the second peak of Y Lliwedd, the rhyolite tuffs and lapilli tuffs are overlain by the basic tuffs of the Bedded Pyroclastic Group.

Descend by the broad, curving ridge from Y Lliwedd towards Gallt y Wenallt. For about 1 km the route is almost along the strike of bedded basic tuffs and lapilli tuffs of the Bedded Pyroclastic Group which generally dip southeast, but then a thick, columnar-jointed, dolerite sill is emplaced along the junction between the Bedded Pyroclastics and a small inlier of bedded Upper Rhyolitic Tuff. It is necessary to keep close to the northerly facing crags to examine the rhyolitic tuffs and lapilli tuffs. Continue along the ridge across further exposures of Bedded Pyroclastics to a second small synclinal inlier of Upper Rhyolitic Tuff just to the south of the ridge leading onto Gallt y Wenallt.

To the north in Cwm Dyli, much of the ground is occupied by a thick dolerite sheet which is well exposed in the crags at Craig Aderyn and Clogwyn Pen-llecheu. The rock is very fresh, uncleaved and gabbroic in texture. On Craig Llyn Teryn the marginal facies of the dolerite is a deep brown-green colour due to the presence of abundant stilpnomelane and actinolite, and is amygdaloidal with siderite and chlorite filling the vesicles.

Regain the track and return along it to Pen-y-pass.

EXCURSION 18. PADARN AND THE PASS OF LLANBERIS

This low-level traverse from Llyn Padarn to Llanberis Pass includes rocks from Arvonian to Caradocian and provides a good stand-by should the weather be bad for one or two days. It is also a traverse which can be managed by those for whom hill walking is not possible.

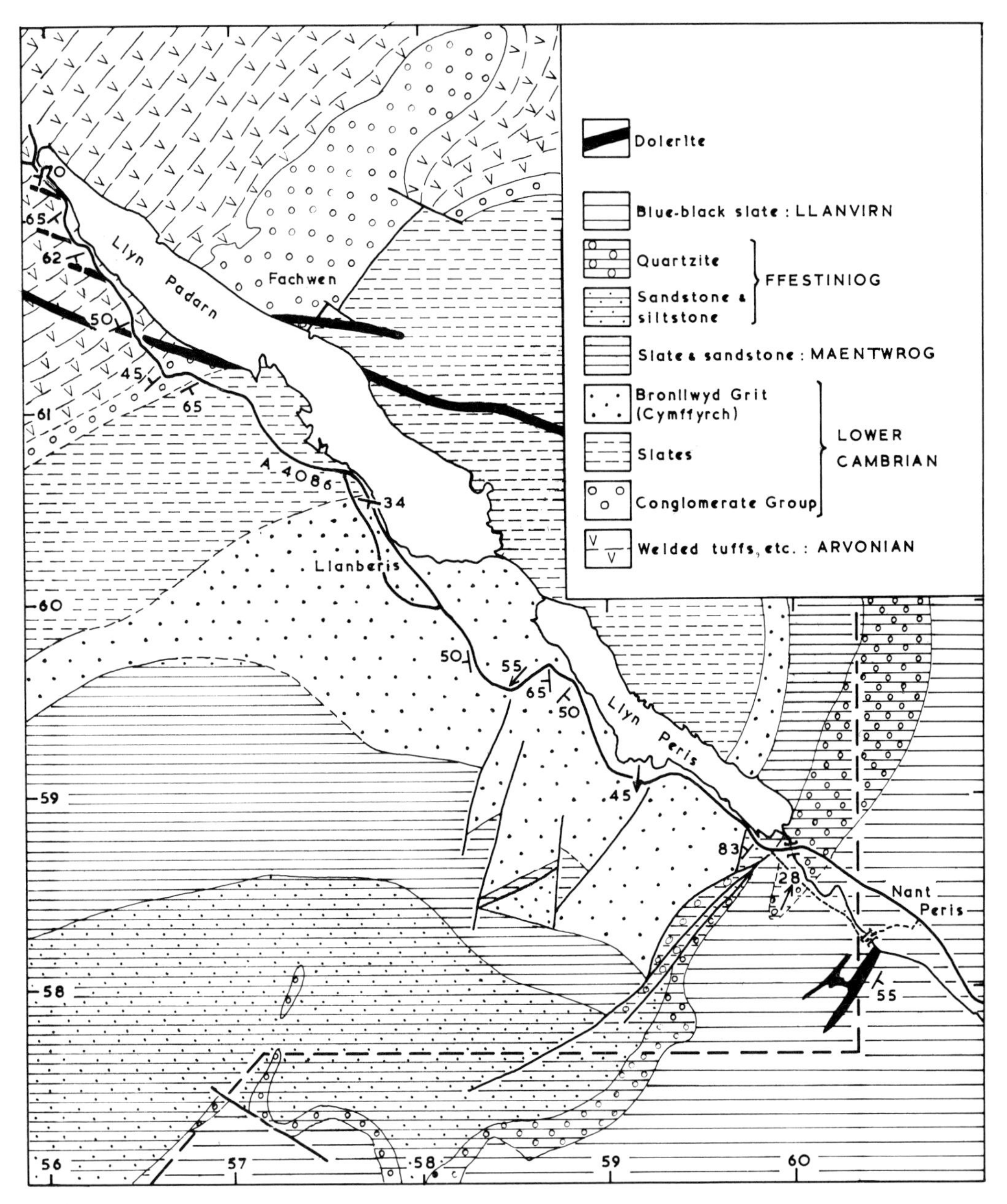

Figure 43. Excursion 18. Partly after Williams (1927) and Wood (1969), with permission.

The excellent section in the Arvonian and basal Cambrian along the lakeside on the Fachwen shore is now not accessible to geologists during the summer, and therefore the itinerary is along the main A4086 road. The succession to be examined is tabulated in figure 4 (column 2) and in figure 5 (column 6). The divisions of the Lower Rhyolitic Group are described on pp 21 and 40.

Use the excursion geological map, figure 43, and the IGS 1:25 000 Geological Special Sheet (Central Snowdonia).
Reference: Williams (1927).
1:50 000 OS sheet No. 115; or 1:25 000 sheets Nos SH55, 56 and 65.

Part 1: Llyn Padarn

The traverse begins at the former road junction at [559623] where, on the southwestern side of the old A4086 greyish green and brown eutaxitic, rhyolitic, welded tuffs of the Arvonian are exposed. The eutaxitic texture is difficult to see because fiamme are very sparse. The rock is fresh and flinty, with phenoclasts of feldspar and corroded quartz. The eutaxitic structure dips 70° towards 284°. At the southern end of this exposure the tuffs are cut by a dolerite dyke.

Walk southeast along the old road where, at the road sign before the right-hand bend, the tuffs are cut by a 20 m wide dolerite dyke inclined very steeply southeast. On the bend the tuffs are seen to be strongly cleaved and to contain clasts of pre-existing welded tuff. Thin quartz/chlorite veins dip steeply northwest parallel to the cleavage. Crystal-rich bands occur within the welded tuffs, which suggests that several distinct flows are present. The banding indicates that the rocks have been deformed and that a minor synclinal hinge is present. Towards the end of the exposure crudely bedded, presumably non-welded tuffs contain blocks of pre-existing welded tuff. Bedding dips 65° towards 142° and hence an anticlinal axial trace has been crossed.

There is then an exposure gap until the point is reached where the old railway track enters a tunnel. Here the road cutting exposes a strongly porphyritic dolerite dyke with feldspars up to 3 cm long. The junction against welded tuffs is vertical and trends at 290°. A chilled marginal facies is developed. Some 10 m beyond the dyke, just before the point at which the road begins to swing to the right, a 45 cm thick lens of tuffaceous slate is interbedded with the welded tuffs. The horizon is partly concealed by gorse and is best seen down at road level; it dips 65° towards 162°. The rocks exposed on the bend itself are green and blue, flinty, welded tuffs but around the bend is another lens of greyish black tuffaceous slate with a strike and dip similar to the first.

The slate is overlain by further welded tuffs until the point is reached at [563618] where a track descends to the road over a built-up wall and incline. Small exposures of sandstone and tuffaceous slate, each about 30 cm thick, occur beneath the incline at the roadside. Bedding dips 75° towards 152°. Welded tuffs overlie the sediments and are exposed for the next 10 m from the end of the incline. The tuffs are then cut by a vertical dolerite dyke trending at 295°. The dyke has a fine-grained marginal facies and it has sent out small stringers into the adjacent tuffs. Xenoliths of welded tuff occur well inside the dyke and, opposite the first lay-by, the dyke is also strongly porphyritic with plagioclases up to 3 cm long. Opposite the southeastern end of the lay-by the dyke is obviously multiple with stringers of chilled, porphyritic dolerite cutting coarse-grained, feebly or non-porphyritic dolerite. Opposite the second lay-by, shear planes cutting the dolerite are coated with coarse-grained calcite, epidote and chlorite.

Welded tuff and further dolerite dykes are exposed and dips continue to be southeasterly until, at the gate to Hafod Wen, the eutaxitic structure in welded tuffs dips northwest and a synclinal axial trace has been crossed. A 5 m thick dolerite dyke cuts the exposures.

Opposite the third lay-by, northwest of the Lake View Hotel at [566612], the road cutting exposes a coarse-grained conglomerate in which the well rounded pebbles are predominantly of welded tuffs and up to 10 cm across. Bedding, indicated by a sandy horizon and a thin slaty lenticle, dips 45° towards 320°. This is the conglomerate horizon which has been taken as the base of the Cambrian succession. It is clear from the traverse thus far that the Arvonian ignimbrites and bedded tuffs pass by interdigitation into a slate, sandstone, conglomerate sequence. The conglomerate is exposed as far as the hotel, where a small exposure gap occurs. The next beds are green and purple sandstones and intensely cleaved, silty sandstones dipping 65° towards 155°. The exposure gap, therefore presumably conceals an anticlinal trace. The silty sandstones contain lenses of coarse-grained, crystal-rich tuffs. These well bedded and laminated rocks are intensely cleaved, and the shear folds are worth study here.

Opposite a long lay-by on the left, the succeeding beds are coarse-grained, green sandstones interbedded with siltstones and slates overlain by green slates. Disharmonic minor folds are present. Opposite the end of the lay-by a minor synclinal hinge is exposed in parallel-laminated, green siltstones and sandstones.

Exposure now becomes poor close to the road but, 100 m further on, there is an exposure of cleaved, massive green sandstone and, immediately beyond this on the eastern side of the road, is an exposure of a massive, greywacke conglomerate containing abundant angular mud pellets. The road now passes across the slate belt but the slates are only poorly exposed near to the road.

Thus purple slates are seen in contact with a sparsely porphyritic dolerite dyke on the right-hand bend just past Glyn Padarn, but there is then a gap until good exposures of purple slate can be examined just before the old road turns off to Llanberis. The slates are folded on the mesoscopic scale and exposures continue to the Bryn Gwyddfor Guest House where they are overlain by a cleaved, green sandstone.

Along the new road the purple slates are overlain by a series of greywacke sandstones up to 2 m thick, with thin slate partings. Sole markings include groove casts and flute casts. The sandstones consist of a massive interval overlain by a parallel-laminated interval, but grading is only occasionally developed. They are probably proximal turbidites and represent the Bronllwyd Grits. Bedding dips 34° towards 190°. The same formation is exposed at the Llanberis Mountain Railway Station, but here mud-pellet greywackes are prominent and the dip is now 50° towards 262°.

The Lower Cambrian succession, from the basal conglomerate upwards, records initially varied shallow-water environments with, perhaps, short-lived reversions to subaerial conditions, but then succeeded by a rapid deepening of the basin and an influx of turbidity-current-borne sediment. Turbidity current action then waned and the background sedimentation led to a predominantly slaty sequence with just occasional greywacke sandstones and conglomerates which represent sporadic turbidity current influxes. Then, once again, renewed turbidity current action led to the accumulation of the proximal turbidites of the Bronllwyd Grit.

Part 2: Llyn Peris and Nant Peris

From a point just south of the Llanberis Mountain Railway Station to the southeastern end of Llyn Peris, new and old road cuttings expose the Bronllwyd Grits. Just beyond the exit from the hotel at [585596] an old road cutting shows massively bedded, coarse-grained greywackes with thin pelitic interbeds folded into a tight anticline plunging at 55° towards 225°. The greywackes have been boudinaged in the limbs of the fold.

At the right-hand bend at [588596] a new road cutting exposes massively bedded, coarse-grained sandstones and fine-grained conglomerates. Mud pellets up to 20 cm across are common. Bedding dips at 65° towards 262°. The greywacke units are up to 2 m thick and consist of a lower massive, sometimes graded interval, followed by a parallel-laminated interval. The beds are the right way up and are cut by a couple of small faults trending at 092°.

The Bronllwyd Grits give way to boulder clay until the lay-by is reached at [592591] and the grits are once more exposed in another new cutting. Here an anticlinal hinge plunging at 45° towards 186° is seen and the complementary syncline to the east is cut and the hinge replaced by a vertical fault trending north–south. Over the next 200 m a fold-pair in well graded greywackes is crossed until at [597588], where a track leads obliquely up to a cottage, well graded, conglomeratic greywackes dip uniformly southeast at about 50°.

75 m before the bend in the road at the end of the lake, greyish black, finely banded slates and siltstones of the Maentwrog Beds are faulted against the Bronllwyd Grit. Bedding dips 83° towards 307°.

It is clear that from immediately north of Llanberis village to the southeastern end of Llyn Peris, the traverse crosses a large, southwesterly plunging fold-pair developed in the common limb of the Arfon Pericline and the Snowdonia Syncline and that minor folds and associated faults are strongly developed within the fold-pair.

Continue along the track instead of the main road and note that the Maentwrog Beds occupy a triangular-shaped area of outcrop on the hillside, bounded by two faults. 200 m along the track the prominent crag on the right is obviously anticlinal (figure 44), the fold plunging at 28° towards 025°. The core is of bedded quartzite which is rich in worm burrows. The quartzite core is overlain on the eastern limb by 8 m of tough, splintery, grey-blue slate; then 15 m of cleaved, impure, fine-

Figure 44. Anticline in Ffestiniog quartzites and silty slates between Nant Peris and the head of Llyn Peris (see text for description).

grained quartzite; then 2 m of blue-grey silty slate; and finally 20 m of cleaved, dirty sandstone which fines upward. The strata are regarded as Ffestiniog Beds and are succeeded, with no obvious intervening fault, by blue-grey slates of Llanvirn lithology exposed in the old quarries.

Return from the quarry to the track and continue across softer Llanvirn slates to the footbridge over the Afon Nant Peris. A prominent, resistant dolerite sill forms a feature striking down the hillside about 100 m upstream from the footbridge. Cross to Nant Peris and walk back along the road until just northwest of the village where the anticline in the Ffestiniog Beds can be seen to advantage.

Part 3: Nant Peris to Pont-y-gromlech

Return across the footbridge and continue upstream on the southeastern bank across the Llanvirn slate and the dolerite sill. The slates contain silty laminae which indicate a dip of 50–60° southeast into the Snowdonia Syncline. The traverse now enters the northwestern margin of the IGS 1:25 000 map of Central Snowdonia.

About 200 m before a prominent spur the soft Llanvirn slates are succeeded, apparently conformably, by tough, splintery, greyish silty slates of the Glanrafon Beds of Caradocian age. A dolerite sill is emplaced along the junction and can be seen striking up the hillside. A second dolerite sill is emplaced within the Glanrafon slates and then the spur itself, which narrows the pass, consists of a well exposed intrusive rhyolite. The rhyolite is very strongly cleaved, but flow-banding and flowage-folding can be found on the ice-smoothed outcrops. The rhyolite was interpreted by Williams (1927) and by Williams (1930) as a series of lavas called the Talgau lavas. On this southwestern side of the valley the intrusive rhyolite has separated into two masses with a thin horizon of sediment between them. Keep just below the brow of the feature where the horizon crops out at [608578]. It consists of a series of lapilli tuffs and sediments, prominent amongst which is a cleaved garnet- and magnetite-rich tuff sufficiently coarsely recrystalline to be called a schist. The garnet–magnetite schist is 2–5 m thick.

Continue across the upper member of the cleaved, intrusive rhyolite and on to exposures of coarse-grained, massive, yet well laminated, sandstones. These are the Gwastadnant Grits of the Glanrafon Beds. Bedding dips consistently at 70° southeast. They are essentially volcanic sands.

The succession thus far from Nant Peris has crossed the soft, blue-black Llanvirn slates which probably accumulated below the reach of wave base under quiet conditions. The succeeding silty, banded, often tuffaceous Glanrafon Slates record an abrupt change to shallower marine conditions and a significant proportion of the detrital fraction is of volcanic origin. The largely volcanic sands of the Gwastadnant Grits indicate a further shoaling and an increase in volcanic activity.

This is a good vantage point to look northeast across the valley to the continuation of the Cwm Idwal Syncline. The core of the syncline in Cwm Patrig is occupied by basic tuffs and basalts of the Bedded Pyroclastic Group. The limbs are formed by the Lower Rhyolitic Tuff with intercalated slates and give rise to the spur of Esgair Felen. The Pitt's Head ignimbrite forms a prominent feature in the northwestern limb (figure 45).

Continue up the succession across the Gwastadnant Grits until a small precipice is reached. Climb around the grits and traverse across the field to a prominent crag in the next field. Cleaved grits are succeeded by a cleaved, bedded, rhyolitic lapilli tuff, then 1 m of cleaved siltstone and then the Pitt's Head formation. The Pitt's Head here begins with alternations of crudely bedded crystal-vitric and lapilli tuffs, all non-welded and strongly cleaved. The tuff loses its bedding, becomes welded upward, and fiamme of flattened pumice are obvious. The tuff contains sporadically distributed clasts up to 60 cm across of pre-existing, feebly welded, rhyolitic tuff. Thus far the Pitt's Head tuffs amount to a thickness of 10 m. At a small rowan tree near the wall a nodular horizon is developed, above which the Pitt's Head becomes intensely welded. This facies is usually taken to be typical of the sheet. Massive columnar jointing is developed. The bulk of the Pitt's Head formation accumulated subaerially.

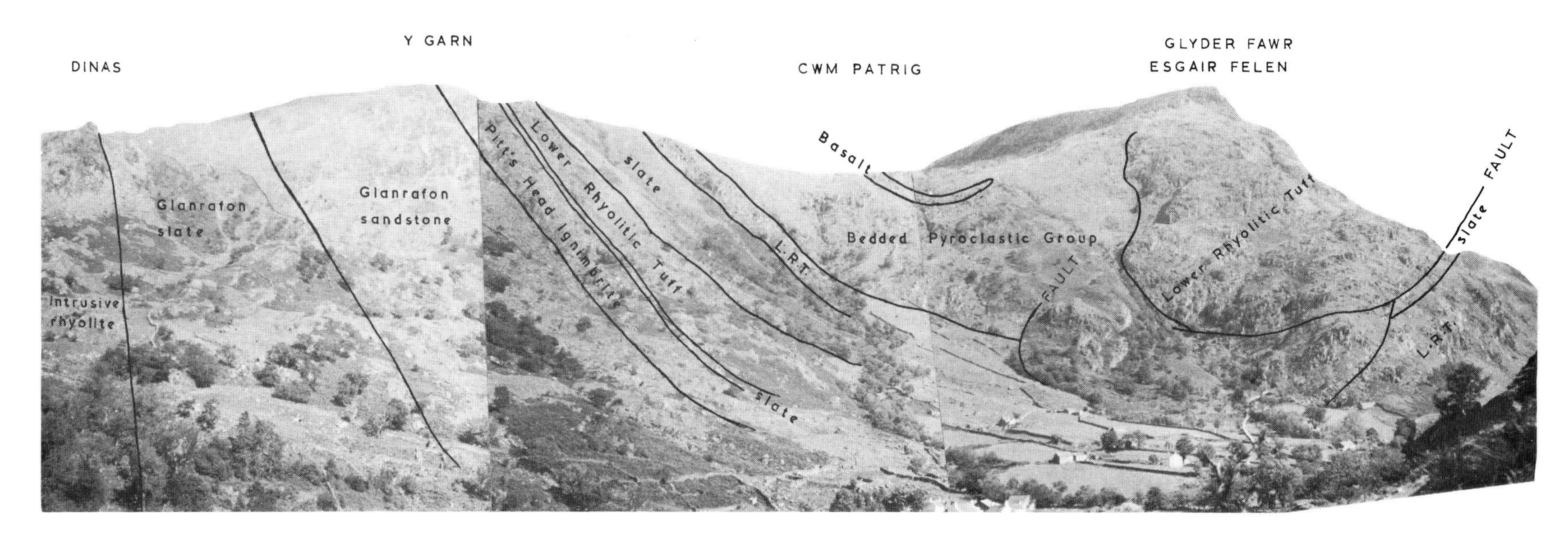

Figure 45. The southwestern continuation of the Cwm Idwal Syncline exposed around Cwm Patrig.

Contour around across a grass- and boulder-covered exposure gap which conceals a small thickness of slate to the next outcrops. These are of cleaved, rhyolitic tuffs of the Lower Rhyolitic Tuff Group, and contain blocks and lapilli of pumice and rhyolite. Bedding cannot be distinguished here, but in about 100 m exposures become well bedded and resemble tuffaceous Gwastadnant Grit. The bedding at first dips 60° southeast but in a futher 20 m becomes first vertical and then slightly overturned to the southeast. A large number of small sill-like rhyolite intrusions are emplaced in the tuffs. Flow-banded and flowage-folded examples can be seen in prominent exposures 200 m above the cottage Cwm Glas-bach.

Make for some obviously well bedded tuffs near a low wall just east of the stream which runs down to the cottage. The tuffs are ripple-marked and show scour-and-fill features and were probably water-lain. A small fold hinge is present here plunging 15° towards 044°. Look up the plunge into Cwm Glas-bach to the asymmetrical syncline, which explains the overturned tuffs seen a little earlier (figure 46). The core of the fold is occupied by the Bedded Pyroclastic Group and the limbs of Lower Rhyolitic Tuff form the summit of Llechog and the spur of Gyrn Las.

Continue across northerly dipping, cross bedded tuffs to the massive, rounded outcrops ahead. These are of flow-banded and flowage-folded intrusive rhyolite. Descend to the road and walk to Pont-y-gromlech.

Part 4: Pont-y-gromlech

A small crag 40 m west of the Pont consists of intensely welded, eutaxitic tuff of the Pitt's Head ignimbrite. The eutaxitic structure dips 48° towards 190° which at this locality probably reflects the attitude of the ignimbrite sheet. Walk southsoutheast over the Pitt's Head along a prominent spur. Near the wall at [631565], coarse-grained sandstones overlie the Pitt's Head. Both parallel-laminated and cross bedded types are present. Bedding dips 55° towards 172° so that the southerly traverse takes one up the succession.

Carry on along the course of the wall to a small crag where columnar-jointed basalts are exposed. They are flow-banded and sparsely vesicular. Overlying the basalts are coarse-grained, cross bedded, slightly calcareous grits which show carious weathering. Horizons rich in brachiopods are present. The sandstones are overlain by cleaved, bedded,

Figure 46. Photograph taken looking southwest into Cwm Glas-bach from the Llanberis Pass. The syncline in the back wall of the cwm verges to the southeast so that beds in the northwestern limb are vertical to overturned.

rhyolitic tuffs and tuffaceous sediments dipping 55° towards 200°. Finally the tuffs are succeeded by a spectacular rhyolitic agglomerate consisting of blocks, bombs and lapilli in a sparse rhyolitic tuff matrix. It shows no internal bedding or grading. The blocks are of sandstone, rhyolite and basalt; the bombs are of vesicular basalt; and the lapilli are of basic pumice, sandstone and rhyolite. This spectacular formation is the Llyn Dinas Breccia, and on this traverse there is no evidence of an unconformity at its base.

Traverse west across the scree fan of large blocks to the very prominent face where the Llyn Dinas Breccia rests on bedded tuffs and tuffaceous sediments dipping 45° south. Continue west across the stile, keeping to the base of the Llyn Dinas Breccias as far as possible. At the apex of the next scree fan the Llyn Dinas Breccia turns over in a fold hinge and cuts the truncated edge of the underlying sandstones and the basalt. The breccia now dips 40° towards 318°, whereas the truncated underlying succession continues to dip 35° south. The base of the breccia strikes stepwise down the hillside and the junction may represent a contemporaneous fault scarp. Whatever the cause, there appears to be a flagrant local unconformity at the base of the Llyn Dinas Breccia.

There is therefore a varied succession of shallow-water, fossiliferous marine sediments, basalt lavas and agglomerates interposed between the Pitt's Head ignimbrite and the Lower Rhyolitic Tuff near Pont-y-gromlech. This contrasts with the succession south of Nant Peris where, as a result of overlap, the Lower Rhyolitic Tuff may rest directly on the Pitt's Head ignimbrite or is separated from it by a thin, discontinuous, slate horizon.

EXCURSION 19. CROESOR TO BEDDGELERT

The traverse is in Caradoc rocks throughout and crosses the Arddu Syncline, the Aberglaslyn Anticline and ends at Beddgelert in an ill-defined and unnamed structure that we might call the Beddgelert Syncline. The traverse includes such items of interest as the composite sills in the eastern limb of the Arddu Syncline, the pumice tuffs of Yr Arddu, a tuff-breccia dyke and the Arddu rhyolite dome, the flagrant unconformity at the base of the Lower Rhyolitic Tuff Group, contact-altered sediments in the aureole above a concealed intrusion, and the postulated Caradocian caldera fault system.

Use the IGS 1:25 000 Geological Special Sheet (Central Snowdonia) as an excursion map.
Reference: Beavon (1963).
1:50 000 OS sheets Nos 124 and 155; or 1:25 000 sheets Nos SH54 and 64.

Part 1: Croesor to Gareg Bengam

A car can usually be parked in Croesor [631447].

Take the track leading northwest up the hill past the chapel to the top of the ridge across Caradocian slates, siltstones and fine-grained sandstones which yield a sparse shelly fauna of Soudleyan age. The siltstones and fine-grained sandstones are commonly parallel-laminated, but some show ripple-drift lamination and bedding planes may be ripple-marked.

At the gate at the top [627451] a feature runs southwest formed by a sill of porphyritic quartz–latite, whereas in the field to the northeast, the feature is absent and a thin, unmapped, non-porphyritic rock is present. A fault is indicated on the 1:50 000 map trending northwest and it occupies a prominent grass-covered depression across the Arddu pumice tuffs.

Continue down the path for 50 m and then go into the field on the right by way of the track. Examine the slates, siltstones and fine-grained sandstones dipping steadily at about 45° northwest. The exposures lie in the southeastern limb of the Arddu Syncline. Carry on down the slope for a further 50 m to prominent outcrops of a composite sill. The rock types present include porphyritic and non-porphyritic quartz–latite and quartz–dolerite.

Carry on down the slope and across the boggy ground to a prominent exposure sticking up from the marshy ground at [624453]. The white-weathering outcrop shows flow-banding, flowage-folding, autobrecciation, and all the characteristics of an intrusive rhyolite. It has been mapped as a microgranite. Some 15 m to the northwest are northwesterly dipping, fossiliferous, massive, cross bedded, coarse-grained sandstones overlying the interbedded slates and siltstones. Immediately to the north across the stream the base of the Arddu pumice tuffs strikes southwest to Gareg Bengam. Walk west to a small sheepfold east of the gate across the track. The sandstones above the sheepfold contain shell bands of poorly preserved, disarticulated brachiopods which have been deformed tectonically. The fauna is of Soudleyan age.

Much of the sand-grade fraction in the coarser sandstones is of volcanic origin: plagioclases and altered lithic fragments of probable andesitic composition are abundant. Generally, the succession westward from Croesor shows an upward increase in grain size and evidence of increased current activity due to shallowing.

Part 2: Gareg Bengam to Yr Arddu

Follow the base of the pumice tuffs northeast for 300 m to the fault lying in the northwest-trending depression. Examine the lowermost pumice tuffs exposed by the wall immediately north of the stream near two rowan trees. It is an unbedded, white-weathering, rhyolitic, pumice-lapilli tuff. The rock is strongly cleaved and chloritised. Pumice lapilli are flattened in the plane of the cleavage. On a fresh surface the rock is in no way unusual: it is a blue-grey, microcrystalline rock with obvious feldspars.

50 m up the western side of the fault hollow from the wall, the tuffs take on a knobbly weathered surface. A near-vertical cleavage is present, but the pumice lapilli now are ellipsoidal and flattened in a plane dipping northwest; in other words, the tuff has a eutaxitic structure parallel to the bottom of the sheet and is welded. 25 m further up the slope, however, the eutaxitic tuffs are overlain by about 15 m of evenly bedded, strongly cleaved, rhyolitic tuffs devoid of pumice lapilli. These are overlain by eutaxitic, massively jointed, welded pumice tuffs. These eutaxitic tuffs continue to a gap at a prominent grass-covered, bedding-plane-like feature seen at an old sheepfold below a ruined wall. Another eutaxitic sheet follows above the gap but the lapilli are less strongly flattened. 10 m above the old wall the eutaxitic tuff is terminated by another prominent bedding-plane-like feature. The volcanics up to this point were grouped together by Beavon (1963) and correlated with the Lower Pitt's Head ignimbrite, but it is clear that several ignimbrite units and bedded units have been lumped together and that the sequence is quite unlike the Pitt's Head ignimbrite in its exposures to the northwest.

A massive lapilli tuff, with rhyolite blocks near the base, succeeds the prominent planar feature. The unit appears to become welded upward and, at the next old sheepfold, eutaxitic tuffs crop out showing thin, continuous horizons 1–2 cm thick, which may represent interbedded vitric tuffs or some kind of secondary modification parallel to the eutaxitic structure. About 15 m above the sheepfold feebly eutaxitic tuffs contain definite interbeds of vitric tuffs 10–15 cm thick, and the rocks hereabouts are strongly cleaved, presumably as a result of their feebly welded and non-welded states. The dip has decreased to about 10° northwest. This second group of volcanics was correlated by Beavon (1963) with the Upper Pitt's Head ignimbrite, but again the group is heterogeneous and is unlike the Upper Pitt's Head ignimbrite in its northwestern exposures.

Cross another bedding-plane-like feature to a sheepfold at the foot of crags of an obvious coarse-grained breccia. The breccia is rhyolitic in composition and has a dyke-like form. The blocks are of flow-banded and folded rhyolite up to 50 cm across and are set in a matrix of rhyolitic lapilli tuff.

The country rock into which the tuff-breccia dyke has been emplaced is correlated by Beavon (1963) with the Lower Rhyolitic Tuff of Snowdon. It contains abundant lapilli and blocks of rhyolite, and pre-existing eutaxitic ignimbrite. The coarse debris is confined to the lowest 5 m and the matrix is feebly welded. The tuff becomes strongly eutaxitic upward and again shows continuous planar features which here seem to be some type of secondary modification because flattened pumice lapilli can sometimes be seen within the bands.

In about a further 50 m the axial trace of the syncline is reached and the beds are almost horizontal. Walk northeast along the axial trace until another prominent, grass-covered depression is reached in front of a high wall. Northeast of the wall, Yr Arddu rises up steeply. About 30–50 m beyond the wall the Arddu rhyolite dome cuts the ignimbrites. Scramble up to the contact and examine the magnificent nodular development at the junction, with nodules up to the size of footballs, though commonly rather less. The nodular shell passes inward to a flow-banded and flowage-folded interior. Look southwest from this vantage point and see the Arddu Syncline spread out in front of you, trending southwest to the alluvium of the Afon Glaslyn.

The process of progressive shoaling apparent in the Soudleyan sedimentary succession beneath the volcanics culminated with uplift and the subaerial emplacement of a series of ignimbrites and lesser quantities of bedded vitric tuff.

Contour northwestward around the hill and, before you strike northwest obliquely downhill to the corner of Coed Caeddafydd and the footpath down to the road, pause and look northwest across Nanmor to Moel y Dyniewyd and Mynydd Llyndy. The hills consist of volcanics of the Lower Rhyolitic Tuff Group and at Moel y Dyniewyd the base of the volcanics is clearly almost horizontal and the steeply inclined sediments beneath can be seen striking northeast up the slope towards the cap of the volcanics.

Part 3: Nanmor to Moel y Dyniewyd

The Nanmor road having been reached, cross the Nanmor river by the footbridge at [621469] and follow the footpath to Coed y Gelli and up onto Craig y Clogwyn. The sandstones exposed in Coed y Gelli at first dip southeast at about 35°. A minor anticlinal axial trace is crossed about 250 m northwest of the footbridge, and for the next 100 m the sandstones dip northwest at about 25°. A complementary minor syncline is then reached and the dip of the sandstones returns to southeast. A minor fold-pair is thus developed in this northwestern limb of the Arddu Syncline.

Make for the wall that runs northwest up towards Moel y Dyniewyd. At the southeastern end of the straight wall, sandstones are faulted against slates and siltstones, and another fault crosses the wall 110 m further northwest up the hillside. These faults were regarded by Beavon (1963) as earlier than the Lower Rhyolitic Tuff in origin and Rast (1969) and Bromley (1969) regarded them as part of the mid-Ordovician caldera fault system. Continue across slates and siltstones dipping at about 35° southeast to some prominent crags of dolerite, the marginal facies of which is strongly porphyritic with plagioclases up to 1–5 cm long in a fine- to medium-grained matrix. From this locality, where the walls join, the columnar jointing of the overlying Lower Rhyolitic Tuff Group is very obvious.

Turn west at the junction of the walls onto exposures of the Llyn Dinas Breccias. The formation in this locality is, in fact, a coarse-grained conglomerate, the pebbles of which consist of well rounded clasts of sandstones, siltstones and occasional dolerites and rhyolites. Cross through the wall at the sheepfold and examine the lower members of the overlying Lower Rhyolitic Tuff. The basal member is an unbedded rhyolitic tuff with lapilli and blocks. The unit becomes nodular upward and is 10–15 m thick. It is overlain by a crystal-rich basic tuff and this, in turn, is followed by a thick eutaxitic and columnar-jointed, welded, rhyolitic, pumice-lapilli tuff.

Traverse west along the base of the Lower Rhyolitic Tuff and observe the overlap of the lens of Llyn Dinas Breccias as the tuffs come to lie on slates and siltstones. At [618476] there is a virtual contact between older slates and siltstones dipping 40° towards 125° and near-horizontal tuffs above providing a superb illustration of the flagrant unconformity of mid-Caradocian age at the base of the volcanic succession.

Descend westward to the marshy ground and cross to exposures on the flank of the spur at about [607474]. Interbedded slates, siltstones and fine-grained sandstones dip 65° towards 172°. The pelitic horizons consist of a mass of pseudomorphs, presumably after cordierite, each up to 5 mm across, and have been interpreted by Beavon (1963) as rocks which have been contact-metamorphosed by a subjacent intrusion.

Continue northwest over the spur where parallel-laminated sandstones crop out and descend into the valley to the west where, at [603476], a small wedge-shaped slice of rhyolitic tuff is faulted down in sandstones dipping 65° south. Traverse due west for 250 m across sandstones which strike east–west and are either vertical or dip steeply south. After 250 m the sandstones are overlain by members of the Lower Rhyolitic Tuff which here again rests with flagrant unconformity upon the older sediments. The surface is irregular in detail but is more or less horizontal. The lowest volcanic rocks are bedded lapilli tuffs, passing up into nodular tuffs, and eventually into eutaxitic welded tuffs. About 75 m to the west, in a gully eroded along a north–south fault, beautifully eutaxitic, columnar-jointed, welded rhyolitic pumice tuffs are exposed. The eutaxitic structure dips west at varying angles up to 60°.

Head due west for the summit of the hill at [597476]. In 75 m the rhyolitic welded tuffs are faulted against members of the Bedded Pyroclastic Group preserved in what is essentially a synclinal structure *in the volcanic sequence*. It is important to realise that because of the flagrant sub-volcanic unconformity there is no correspondence in this area between structures in the sedimentary rocks and structures in the overlying volcanic sequence. The lowest member of the Bedded Pyroclastic Group is a basaltic agglomerate consisting of basalt blocks with basalt and rhyolite lapilli in a matrix of basaltic tuff. This pyroclastic horizon was correctly mapped by Beavon (1963), but has been omitted from the IGS 1:25 000 geological map. This is overlain by a strongly cleaved, flow-banded, flowage-folded, autobrecciated vesicular and amygdaloidal basalt. This rock persists almost to the summit where, having crossed the synclinal axial trace, it is underlain by what is essentially a rhyolitic agglomerate and agglomeratic tuff. Bedding dips 75° towards 143°; cleavage strikes at 038° and is vertical. Both the IGS map and Beavon (1963) record the summit as basalt.

From the summit descend north and then northwest across basalt and basaltic tuff in the northwestern limb of the small syncline, then across an anticlinal structure in cleaved rhyolitic tuffs of the Lower Rhyolitic Tuff Group. As the slope steepens, the largely grass-covered surface practically coincides with the northwestern dip of the bedded rhyolitic tuffs in the northwestern limb of the anticlinal structure.

Pass through an iron gate in a prominent wall, and bedded lapilli tuffs of basaltic composition are seen to overlie the rhyolitic tuffs and occupy another small synclinal structure. The final slope down to Beddgelert is in rhyolitic tuffs of the Lower Rhyolitic Tuff Group in the northwestern limb of the small synclinal structure. These rocks are eutaxitic tuffs, bedded crystal-vitric tuffs and lapilli tuffs dipping at about 50° due east.

Return to Beddgelert by the footbridge over the river.

Ardudwy

EXCURSION 20. TAN-Y-GRISIAU AND MOELWYN

The aim of this excursion is to examine the Tan-y-grisiau granite and its metamorphic aureole, and to examine the Tremadoc to Caradoc succession as developed in the Moelwyn Hills. The Stwlan hydro-electric scheme is also worth a visit in passing.

Use the IGS 1:25 000 Geological Special Sheet (Central Snowdonia) as an excursion map.
Reference: Bromley (1969).
1:50 000 OS sheets Nos 115 and 124; or 1:25 000 sheet No. SH64.

Locality 1: [689443]

Park in the large lay-by at the highest point on the B4414 and examine the extensive, fairly new road cuttings on the east.

Much of the rock exposed at the northern end of the cutting is a medium-grained, pale pink-weathering granite with many chlorite-rich patchy segregations. To the south is a deeply weathered zone which is cavernous and rich in hydrated ferric oxides. Recognisable quartz–chlorite veins cut the weathered zone, and the whole represents a leached and oxidised, formerly sulphide-rich zone. Less completely leached and oxidised pyrite-rich belts remain: these are nearly vertical and strike at 093°. On the western side of the road at the lay-by, a natural pavement shows the granite cut by a basic facies (or it may be an irregular dolerite dyke) which, in turn, is cut by quartz–chlorite veining. The basic lenses are vertical and trend at 072°.

Locality 2: Tan-y-grisiau Quarry [694453]

A car can be left at the start of the track leading to the quarry. The quarry is now being rapidly in-filled with the wrecks of cars and other refuse.

The exposed rock is a grey-green microgranite with chlorite-rich clots 0·5–1 cm across and coarse-grained xenoliths, each with a whitish reaction rim. The rock consists of plagioclase (albite–oligoclase), perthite, quartz and chlorite after biotite. Magnetite, zircon and allanite are common accessories. The upper contact with the Tremadoc siltstones and slates is well exposed towards the top of the quarry face and a distinctive roof facies is developed. It is fine-grained and vesicular; the perthite has been converted to muscovite, and the plagioclase to an aggregate of albite, quartz and calcite. The changes are presumably autometasomatic.

Although the contact is discordant in detail (figure 47), the intrusion as a whole has the form 'of a truncated ellipsoidal cone with the cone axis inclined about 60° to the northwest' (Bromley 1969). The country rock above the intrusion is obviously contact-altered: it is heavily spotted, the spots consisting of chlorite and sericite after cordierite, and occasionally after andalusite. The assemblage is now largely albite–epidote–chlorite–sericite–quartz, but clearly the cordierite porphyroblasts indicate that the hornfelses originally belonged to the hornblende hornfels facies close to the contact but were subsequently retrogressed to the albite–epidote hornfels facies, probably as a result of the explusion of volatiles at a late state in the crystallisation history of the granite.

There is another small quarry to the left of and below the main quarry, which has a fence across the top of it. On the northeastern face of this quarry there are vertical pipe-like masses of coarse-grained calcite

Figure 47. The face of Tan-y-grisiau quarry, showing the discordant nature of the roof of the granite intrusion.

carrying allanite, pyrophyllite, quartz, and traces of molybdenite. **These exposures should not be hammered.** The pipes are presumably the result of late-stage hydrothermal activity.

Examine the hornfelses near the main road. Note the extent to which the contact metamorphism has emphasised the bedding. Most of the porphyroblasts are sac-like in shape and were originally cordierite, but occasionally rectangular shapes suggest former andalusite. Notice that cleavage is present as a distinct fabric. Bedding dips 25° towards 360°; cleavage dips 75° towards 342°.

Locality 3: Traverse from Tan-y-grisiau to Stwlan Dam

Leave your transport in the free car park at the CEGB Exhibition Hall at [683449]. Take the road leading northeast.

Examine the exposures of porphyroblastic hornfels. Notice the scour-and-fill structures in these banded and laminated Tremadoc siltstones and pelites. The bedding dips 35° towards 330°, so that the traverse takes us across an ascending sequence.

Continue around a tight left-hand bend in the road. Continue west along the road to the point where it again bends, this time gently towards the northwest. Leave the road at this point and head west for 50 m to a small crag with a small oak tree and a few hazel bushes on it. The rock exposed is grey-weathering and weathers with a rounded rather than flaggy surface, as did the underlying Tremadoc siltstones and pelites. The rock is still spotted and is a massive, dirty sandstone at the base passing up into a laminated sandstone. About 3 m is exposed and it represents the basal Arenig sandstone.

Cross northeast to the exposures near the waterfall. These rocks overlie the basal sandstone and are Arenig flagstones. They consist of laminae and lenses of fine-grained sandstone alternating with silty mudstone. Lichen-free surfaces are difficult to find but they show the rocks to be strongly burrowed.

Return to the road. 40 m further along from the point at which the footpath from the waterfall joins the road, a dolerite dyke can be seen cutting the Arenig flags. The dyke is 2 m wide at the top and widens downward to 10 m.

Carry on up the road to an exposure 10 m southeast of the point at which the prominent incline is crossed by the road. The Arenig flags are strongly burrowed, and it is noticeable that the effects of contact metamorphism are weaker. Look up the incline to the free face formed by a sill of quartz–latite emplaced along the junction of the Arenig flags and the overlying Moelwyn Volcanic Group.

A new cutting on the right, 15 m up the road from the incline, exposes strongly burrowed, cross bedded and ripple-marked siltstones. Continue up the road beyond the point at which the gradient slackens a little. Just before some huge boulders at the foot of a prominent crag of rhyolite is reached, soft, blue-grey, finely cleaved slates are exposed overlying the flaggy beds. These beds are lithologically similar to much of the Llanvirn slates of central and western Snowdonia. About 100 m further up the road the soft slates are seen to be about 20 m thick and underlain by interbedded pyritous sandstones and slates with a couple of thin vitric tuffs. The tuffs and some of the sandstones have developed load casts into the slates beneath. All the beds are strongly burrowed.

A little further up the road the slates have been oxidised so that they are very ochreous. Fresh pieces contain horizons which are a mass of pyrite cubes.

In a short distance, at [674448], a prominent fault occupying a gully throws the base of the rhyolite sill back from the road. Carry on for 500 m to the second bend of a double hairpin. The Arenig flagstones exposed in the road cutting carry a good vein of quartz and galena. Notice that the sediments are now unaffected by contact metamorphism.

Carry on for 150 m to a right-hand bend at [670444] where a prominent slot in the cliff is occupied by a 10 m thick dolerite dyke. After following the left-hand bend in the road, go through a gap in the wall and examine the contact of the quartz–latite sill with the underlying slates. There is a thin, impersistent basal facies 0–50 cm thick consisting of flinty microcrystalline rhyolite. This is overlain by 2–3 m of autobrecciated quartz–latite and above this, the rock is flow-banded and flowage-folded.

At the next bend the road crosses the base of the quartz–latite sill and an excellent section through it is exposed from this point up through the hairpin bends to beyond the dam. It is seen to consist of zones of autobreccia alternating with zones of flow-banded and flowage-folded quartz–latite.

From the end of the dam climb up northeast and then east along the top of the sill to a prominent perched block at the head of a feature eroded along the roof of the sill. There are excellent exposures of bedded rhyolitic tuffs of the Moelwyn Volcanic Group on the northern side of the feature, but the base of the volcanic succession is, in fact, about 4–5 m down the scarp face to the south. The base consists of 1 m of crudely bedded agglomeratic rhyolitic pumice tuff. This is overlain by 1 m of well bedded and fine-grained lithic tuff with just occasional lapilli, followed abruptly by 2 m of agglomerate with rhyolite blocks up to 30 cm across. The agglomerate becomes crudely bedded upward as grain size decreases, and it passes into a lapilli tuff, which shows irregular, hummocky bedding and is 7 m thick. It is overlain by 1 m of finely laminated vitric tuff and then by about 7 m of parallel-laminated tuff in the grassy gully. The excellent exposures on the northern side of the feature are in crudely lenticularly bedded tuffs and tuffaceous sediments and, 20 m down the grassy gully, these have been thrown into very large folds which have the characteristics of soft-sediment deformation. The disturbed beds are overlain by 30 m of unbedded rhyolitic agglomerate with some blocks over 1 m across. It may be that the soft-sediment deformation in the underlying tuffs and sediments was caused by loading due to the sudden emplacement of the agglomerates, or they may be slump folds. The agglomerate fines upward and becomes crudely bedded towards the top.

Proceed northwest across the strike. There is then a 15 m exposure gap before 5 m of laminated silty slates, siltstones and fine-grained sandstones crop out. These are mapped as Glanrafon (Caradoc) sediments.

The sediments are succeeded by another rhyolitic agglomerate, which is rather finer-grained in that the blocks range up to 30 cm across. The unit fines upward to a uniform, crudely bedded, lapilli tuff. There is then an exposure gap of 60 m. Mapping indicates the ground is occupied by higher members of the Moelwyn Volcanic Group.

The next exposures are of grey, blocky, columnar-jointed felsite: another quartz–latite sill. The rock is flow-banded, the bands being parallel to the contact near the base (and the columnar jointing is normal to this structure), but 40 m above the base the rock is strongly flowage-folded. Just beneath the old tramway a zone of autobreccia is exposed, but this returns to a flow-banded and folded facies except for the topmost 30 m, which is strongly autobrecciated.

The roof of the sill dips northwest towards a pair of old slate quarries. Walk down the roof of flow-banded facies and across a prominent quartz vein about 50 cm wide onto the peat-covered Glanrafon slates. The eastern quarry exposes silty slates and fine-grained sandstones and a prominent white-weathering vitric tuff which, at the northwestern end of the quarry, is agglomeratic. The slates and siltstones are generally parallel-laminated and bedding dips at about 30° northwest. Sediments of this type persist, strengthened by dolerite and composite sills, to the Snowdon Volcanic Series of the Arddu Syncline.

Arllechwedd

EXCURSION 21. CARNEDD Y FILIAST TO CWM IDWAL

The aim of this excursion is to examine the sequence from the Cambrian Slate Belt to the core of the Cwm Idwal Syncline. The excursion involves about 12 km of very pleasant hill-walking. The sequence is tabulated in figure 4 (column 2) and figure 5 (column 6).

Use the excursion geological map, figure 48 (route A), and the IGS 1:25 000 Geological Special Sheet (Central Snowdonia).
Reference: Williams (1930).
1:50 000 OS sheet No. 115; or 1:25 000 sheets Nos SH65 and 66.

Part 1: Y Fronllwyd to Mynydd Perfedd

A car can usually be left on the verge of the road near the road junction at [630637]. Take the track northwest from this point for about 400 m and then turn west and scramble up the ridge of Y Fronllwyd, the type locality of the Bronllwyd Grits.

The rocks consist of a series of greywacke sandstones, often conglomeratic, with thin interbeds of mudstone. Sandstones are usually about 1 m thick and may show a graded interval, followed by a parallel-laminated interval, and sometimes then by a cross laminated interval. Convolute lamination and slumped laminae are common and the beds are probably proximal turbidites.

The increase in the width of outcrop as the formation is traced southwest up the ridge can be shown to be due largely to the presence of a minor fold-pair plunging at 30° towards 048°.

Just before the grass-covered saddle at the head of Cwm Ceunant, southeasterly dipping, grey and green, silty slates with thin, fine-grained sandstones crop out. They are well exposed in the cliff at the head of the cwm. These Maentwrog slates and fine-grained sandstones coarsen upward and can be examined on the steeper slope up to the summit of Carnedd y Filiast. Greywacke sandstones increase in number, grain size and thickness and show grading, cross lamination, slumping and convolute lamination. Sole markings include flute casts, load casts and groove casts. Irregular burrows are common. Thus turbidity current action became increasingly important towards the upper part of the Maentwrog stage.

Continue to the prominent crags of cross bedded Ffestiniog quartzite just below the summit. The beds dip southeast into the Snowdonia Syncline. From the summit look across to the magnificent exposures of Bronllwyd Grits at Marchlyn Bach and Maentwrog and Ffestiniog Beds at Marchlyn Mawr. (Access to these exposures is difficult during the present construction work.) See figure 49. The Ffestiniog Beds on the summit are plane-bedded quartz-granule conglomerates with abundant burrows.

Cross the wall by the ladder to the top of Cwm Graianog. The southeasterly dipping bedding planes of the Ffestiniog quartzites and conglomerates are superbly rippled and include straight-crested, symmetrical, asymmetrical and flat-topped types, as well as interference ripples. If time permits, descend part of the way into Cwm Graianog along its northwestern wall and examine the different types of ripples and cross beds. The trace fossil *Cruziana semiplicata* is common and *Rusophycus, Phycodes, Diplichnites, Dimorphichnus* and *Planolites* also occur. The sediments were deposited above wave base but below low water mark, where the bottom was swept by periodic currents.

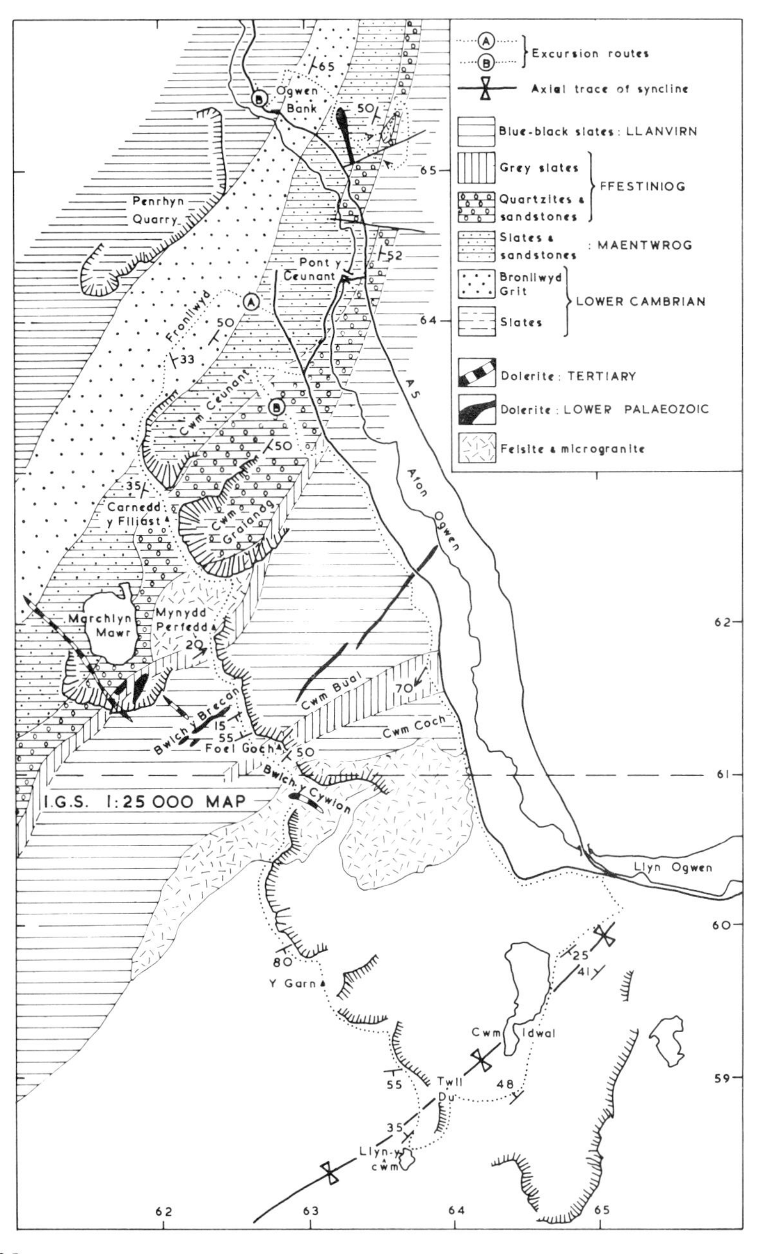

Figure 48. Excursion 21. Largely after Williams (1930) and Evans (1928), with permission.

Return to the top and carry on around the head of the cwm, the southern wall of which is formed by the Mynydd (Moel) Perfedd granophyre. This is a swollen, sill-like mass. It is a grey, medium-grained, feldspar-porphyritic rock. Specks of chlorite after biotite, often aggregated in clots, can be readily distinguished in the base of quartz and feldspar.

Part 2: Mynydd Perfedd to Y Garn

Continue to the summit of Mynydd Perfedd where the southern junction of the sill has been mapped as a fault. The succeeding beds are exposed on the slope down to Bwlch y Brecan. They are tough, splintery, grey-blue slates with thin silty laminae which Williams (1930) has mapped as Llanvirn but, since no fossils have been obtained from them, their age is not known. The beds are tightly folded on a mesoscopic scale and folds plunge at 20° towards 055°.

Immediately below the crags of resistant silty slates, in the bwlch, soft blue-black slates of typical Llanvirn lithology crop out. Away to the southwest, in Cwm Dudodyn, on the same strike, Llanvirn graptolites have been obtained by Williams. An outcrop of a quartz–dolerite sill is seen 150 m southwest of the bwlch at [623614]. It is a grey-brown, strongly weathered and cleaved rock.

Continue along the top of the back wall of Cwm Bual to the summit of Foel Goch where grey silty slates, occupying the core of an anticlinal fold and mapped as Llanvirn by Williams, crop out. They may represent an inlier of Ffestiniog Beds. Dip measurements on the way up Foel Goch indicate the presence of minor folds plunging at low angles towards 050°.

200 m to the southeast of the summit of Foel Goch, the silty slates are overlain by the marginal facies of the Bwlch y Cywion microgranite sill-like intrusion. The marginal facies is an autobrecciated felsite but it passes inward to a white or pink feldspar-porphyritic microgranite in the matrix of which small flakes of chloritised biotite can be distinguished. The route now enters ground covered by the IGS 1:25 000 Geological Special Sheet (Central Snowdonia).

Figure 49. View from Carnedd y Filiast looking southwest towards Elidir Fawr. This magnificent section from the Cambrian slates to the Llanvirn slates will be accessible once more upon the completion of the Dinorwic hydro-electric storage scheme.

Continue due south for another 150 m down the steep slope to a stream which has followed a readily eroded olivine–dolerite dyke of Tertiary age. The poorly exposed dolerite is spheroidally weathered.

Return to the top of the cwm and continue across the microgranite to its upper contact where again it is an autobrecciated felsite, much cut by quartz veins. The contact dips southeast and the quartz veins also penetrate the overlying sediments. The sediments are mildly contact-metamorphosed, Caradocian silty slates of the Glanrafon Beds.

Continue across the bwlch and on the way up towards Y Garn the slate screes are worth searching for fairly common biserial graptolites. In another 75 m a coarse-grained dolerite sill is crossed. The sill is followed by further slates and thin basic tuffs and agglomeratic tuffs and then an unmapped, cleaved, vesicular dolerite. Bedded rhyolitic tuffs follow. These are distinctively cream-weathering, strongly cleaved rocks and the cleavage has been subsequently deformed by a well developed set of kink bands. The bedded basic tuffs, rhyolitic tuffs and interbedded sediments must lie close to the horizon of the Carnedd Llywelyn Volcanic Group. The rhyolitic tuffs are followed by a series of interbedded sandstones, slates and mud-pellet sandstones which contain large cubes of pyrite and fairly common, tectonically distorted brachiopods. The sediments constitute the Glanrafon sandstones.

On the fairly prominent flat, before the summit of Y Garn, the traverse crosses another coarse rhyolitic tuff which, in places, is agglomeratic and is a distal representative of the Capel Curig Volcanic Group. The succeeding sandstones are highly fossiliferous and brachiopods are easily obtained from blocks in the scree. The beds dip consistently to the southeast and lie in the northwestern limb of the Cwm Idwal Syncline.

The summit of Y Garn consists of the Pitt's Head ignimbrite. The sheet has a basal zone of non-welded pumice-lapilli tuff of rhyolitic composition which passes up into a welded zone where flattened pumice lapilli impart a eutaxitic structure. The degree of welding decreases towards the top where a nodular horizon is developed and the rock is strongly cleaved.

Part 3: Y Garn to Cwm Idwal

Look southeast towards Llyn-y-Cŵn. The axial trace of the Cwm Idwal Syncline runs northeast–southwest about 200 m this side of Llyn-y-Cŵn.

Descend from Y Garn towards the top of Twll Du across slates, siltstones and sandstones containing occasional, often fragmentary, shelly fossils and then a thin lens of agglomeratic, rhyolitic, lapilli tuff. Further sediments follow and fossils, especially coarse-ribbed orthids, are readily found in the loose blocks. The fossiliferous beds are followed by the Lower Rhyolitic Tuff Group which begins with cleaved, unbedded, rhyolitic, crystal-vitric tuffs. An unmapped intrusive rhyolite, which is autobrecciated, flow-banded, flowage-folded and columnar-jointed, is emplaced in the vitric tuffs and forms a small buttress in the cliff a few metres above the base of the tuffs.

Continue the descent across crystal-vitric tuffs to an intercalation of slates, siltstones and sandstones. The sediments are followed by massively jointed, rhyolitic, pumice-lapilli tuffs in which bedding cannot be distinguished. The rocks appear to be non-welded and may represent sillars. The massive tuffs are overlain by bedded lapilli tuffs, and the highest beds in the group are flinty vitric tuffs interbedded with tuffaceous siltstones in which brachiopods are common.

The Lower Rhyolitic Group is succeeded by the Bedded Pyroclastic Group. The group begins with obviously bedded, green, basic tuffs, lapilli tuffs and agglomeratic tuffs. The lower beds are finer-grained, and are often cross laminated and ripple-marked. Clasts are commonly of basic pumice and basalt, but rhyolite clasts are also present. Trilobites and brachiopods occur in some of the finer-grained horizons and clearly the rocks were water-lain. The tuffs are succeeded by basic lavas which are generally very rubbly and individual flows are very difficult to distinguish. The rocks are often amygdaloidal and feldspar-porphyritic, and towards the top show a crude development of small pillows. The lavas were therefore also sub-marine. It is perfectly evident that the basic rocks occupy the core of a syncline and the youngest rocks seen are further lapilli tuffs and agglomeratic tuffs which can be seen resting on the lavas about 100 m northwest of Llyn-y-Cŵn.

From Llyn-y-Cŵn take the footpath which leads down the southern side of the Twll Du into Cwm Idwal. The path lies practically along the junction of the Bedded Pyroclastic Group with the underlying Lower Rhyolitic Tuff Group in the southeastern limb of the Cwm Idwal Syncline. Take the path from the bottom of Twll Du that leads east and then north around Cwm Idwal. The traverse crosses largely scree-covered, columnar-jointed, flinty vitric tuffs of the Lower Rhyolitic Tuff Group and then a stream which has cut its course along the bedding planes of sandstones, siltstones and slates underlying the flinty tuff.

As the Idwal slabs are approached, notice the prominent, more or less horizontal, *en echelon*, quartz-filled tension gashes cutting the Lower Rhyolitic Tuff. The rock is mainly a massive, crudely bedded, pumice-lapilli tuff. Some of the thicker beds may be feebly welded and it is possible that the bulk of the rocks are sillars and, if so, would have been emplaced subaerially.

The path along the eastern side of Llyn Idwal lies in the Lower Rhyolitic Tuff and, about 100 m before a wall at [647596], crosses the axial trace of the syncline. At the northern end of the lake the path turns northeast along the strike close to the base of the tuffs. It then crosses onto Glanrafon sandstones and curves down to Ogwen Cottage, immediately south of which it crosses the Pitt's Head ignimbrite and so reaches the A5 road.

Return to your transport by way of the old road.

EXCURSION 22. CWM IDWAL TO CAPEL CURIG

This excursion covers ground in which major structures—the Cwm Idwal Syncline and the Cwm Tryfan Anticline—have developed within the maximal structure of the Snowdonia Syncline. It continues the traverse of the previous excursion.

The references and the succession are the same as for the previous excursion, except that 1:25 000 sheets Nos SH 65, 66 and 75 may be used as alternatives to the 1:50 000 OS sheet No. 115.

Use the IGS 1:25 000 Geological Special Sheet (Central Snowdonia) as an excursion map.

Locality 1: Cwm Idwal

Take the footpath leading south and then southeast from Ogwen Cottage across the Pitt's Head ignimbrite and onto sandstones and siltstones above.

The Pitt's Head ignimbrite has a cleaved, non-welded upper facies which passes down into a eutaxitic, strongly welded tuff. The eutaxitic structure dips westward into the syncline and the sheet is cut by quartz-filled tension gashes. Continue along the path, through a 90° turn and so southwest onto the overlying rhyolitic rocks. At this point the Lower Rhyolitic Tuff overlaps the Llyn Dinas Breccias. The breccias thicken up to the southeast and can be seen to consist of blocks of sandstone and rhyolite in a matrix of rhyolitic tuff. The locality lies more or less on the axial trace of the Cwm Idwal Syncline.

At the northern end of Llyn Idwal view the Cwm Idwal Syncline, the axial trace of which is outlined in section by the cleft of Twll Du (figure 50). Cross to the small beach and on to a prominent knoll of the Pitt's Head ignimbrite. The rock is strongly eutaxitic and thin quartzo-feldspathic lenses are often 30 cm long. The exposure presents some interesting features. Cross bedded Glanrafon sandstones are overlain by highly contorted siltstones and fine-grained sandstones. These are overlain by a feebly eutaxitic welded tuff, then a raft of strongly contorted sandstones and, finally, strongly eutaxitic ignimbrite. It would therefore seem that the underlying sediments were capable of undergoing soft-sediment deformation and were wet at the time of emplacement of the ignimbrite in this locality. The exposures now lie in the northwestern limb of the Cwm Idwal Syncline.

Continue south along the western shore of the lake to a glaciated pavement of the Lower Rhyolitic Tuff Group. The group consists mainly of unbedded, cleaved, rhyolitic, pumice-lapilli tuffs. A few beds of crystal-vitric and crystal-lithic tuffs 20–50 cm thick are included.

Cross ground largely covered by hummocks of morainic debris beyond which are exposed sandstones and siltstones interbedded within the Lower Rhyolitic Tuff. The overlying tuffs, exposed where the track becomes steep, are finely bedded and laminated vitric tuffs of flinty

Figure 50. View of the Cwm Idwal Syncline from northwest of Llyn Ogwen, which forms the middle foreground. The rock units indicated are (1) Glanrafon sandstones; (2) Pitt's Head ignimbrite; (3) sandstones above the Pitt's Head ignimbrite; (3a) a pair of unnamed rhyolitic tuffs in the southeastern limb; (3b) overlying sandstones; (4) Lower Rhyolitic Tuff; (4a) slates, siltstones and sandstones within the Lower Rhyolitic Tuff; (5) tuffs and basic lavas of the Bedded Pyroclastic Group.

appearance. They are overlain by another intercalation of sandstones and siltstones, which show intense contemporaneous soft-sediment deformation, and in turn by dark blue, massive, rhyolitic, vitric tuffs which show a crude columnar jointing. These massive tuffs pass up into finely columnar-jointed tuffs very well seen to the east of Twll Du.

Cross the axial trace of the syncline into the southeastern limb and ascend the back wall of the cwm by way of the track more or less along the base of the Bedded Pyroclastic Group.

Locality 2: Llyn-y-Cŵn to Y Glyder Fach

From the top of the cwm cross to the southeastern shore of Llyn-y-Cŵn where the junction of the Bedded Pyroclastic Group with the underlying Lower Rhyolitic Tuff Group is exposed.

Climb up towards Glyder Fawr by way of the small stream immediately northeast of Llyn-y-Cŵn across rhyolitic lapilli tuffs with occasional bombs of rhyolite. Notice the way in which the lapilli tuffs were deformed contemporaneously beneath the rhyolite blocks and bombs as a result of impact. The northwesterly dip of the rhyolitic pyroclastics is steeper than the slope of the hillside so that, as the hill is ascended, the succession is gradually descended. The underlying beds are beautifully columnar-jointed, structureless, flinty vitric tuffs and then, about 200 m above Llyn-y-Cŵn, a flow-banded, flowage-folded and autobrecciated, intrusive rhyolite crops out. The IGS map indicates that the same rock crops out from this point to the summit of Glyder Fawr, but much of the rock exposed northwest of the summit of Glyder Fawr, and indeed at the summit itself, has more the field characteristics of a cleaved, agglomeratic, lapilli tuff of rhyolitic composition. The clasts, flattened in the plane of cleavage, consist mainly of rhyolite but clasts of pumice, sandstone and basalt are also present. Bedding can be found and close to the summit it dips 52° towards 322° into the Cwm Idwal Syncline. Flow-banded and flowage-folded rhyolite does crop out however, south and west of the summit. Quite a lot of unmapped vesicular basalt (or intrusive dolerite?) is also present within the rhyolitic tuffs around about the summit.

Walk northwest from the summit across unmapped rhyolitic tuffs and vesicular basalts (or dolerites) and occasional intrusive rhyolites to a massive dolerite which occurs, as the 1:25 000 map indicates. Cross the pavement of frost-heaved dolerite blocks to the northeastern margin of the intrusion, where it cuts vesicular, blocky basalt lava flows. Walk southeast around the top of the back wall of Clogwyn Du across the basalts onto sandstones and then an apophysis of the dolerite intrusion once more cuts sandstones with thin slate interbeds belonging to the Glanrafon sandstones.

Continue around the top of the cwm to the Pitt's Head ignimbrite dipping northwest into the Cwm Idwal Syncline. The top of the sheet is an unwelded pumice-lapilli tuff which becomes welded with a prominent eutaxitic structure downwards. The fiamme are deformed lapilli of chloritised pumice. The degree of welding increases downward until a nodular horizon below which the welding gradually decreases until the base is reached. The sheet shows good columnar jointing and can be seen striking obliquely across the ridge of Y Gribin down to and across the centre of Llyn Bochlwyd.

Continue east across Bwlch Ddwy Glyder, where Glanrafon sandstones and slates are exposed beneath the Pitt's Head ignimbrite, and on to the Capel Curig Volcanic Group. The youngest members of the group are bedded, rhyolitic, pumice-lapilli tuffs. The tuffs are well cleaved and lapilli are strongly flattened in the plane of cleavage. Climb up towards the summit of Y Glyder Fach across a thin, flow-banded and flowage-folded rhyolite sill and on to the agglomeratic lapilli tuffs of the Glyder Fach Breccias. The blocks are of rhyolite and bedded tuff in a matrix of pumice-lapilli tuff. The Glyder Fach Breccias form a thick wedge, as the IGS map clearly indicates. It seems very likely that the poorly exposed southern slopes of Glyder Fach conceal the source vent of the Capel Curig Volcanic Group.

Locality 3: Glyder Fach to Capel Curig

Walk northeast from the summit of Y Glyder Fach to the top of the Bristly Ridge which leads down to Bwlch Tryfan. The crest of the ridge is formed by a northwesterly dipping sheet of flow-banded and flowage-folded rhyolite—the same intrusive sheet that forms much of the steep eastern face of Tryfan. Beneath the rhyolite sill, a further thickness of the Glyder Fach Breccia crops out.

Figure 51. View of the Cwm Tryfan Anticline from 2·5 km northeast of Tryfan. The section illustrated is 3 km in length. The approximate trace of the anticline and the more important lithological formations are shown.

Continue southeast around the top of Cwm Tryfan on the descent towards Llyn Caseg-fraith. The next 500 m lies across another flow-banded and flowage-folded, sill-like rhyolite mass with a thin raft of sandstones. The sandstones dip at 25° to the southwest, and the swing in strike indicates that the traverse is approaching the closure of the periclinal Cwm Tryfan Anticline (figure 51).

Continue east for 200 m across intrusive rhyolite and on to southeasterly dipping Glanrafon sandstones in the hinge zone of the pericline. The sandstones are magnificently exposed in the wall of Cwm Tryfan. They are highly tuffaceous with abundant feldspars and pumice lapilli that tend to weather out to give cavities, and they are often well graded. Washouts and trough cross bedding are common, and laminae up to 1 cm in thickness are rich in heavy minerals. These essentially volcanic sands accumulated in shallow water under the influence of strong currents.

Immediately west of Llyn Caseg-fraith, a northwest-trending, spheroidally weathered, Tertiary olivine–dolerite dyke is poorly exposed. The sandstones here dip south and the axial trace of the Cwm Tryfan Anticline has been crossed.

Continue around the head of the cliffs to the unnamed summit at [677582]. Notice that the sandstones now dip uniformly southeast. A sill of flow-banded and flowage-folded rhyolite trends northeast across the summit and immediately to the southeast of the summit a raft of sandstone is seen within the rhyolite. Follow the strike northeast to the summit at [685586] and examine the outcrops of the Capel Curig Volcanic Group. Much of the tuff is massive, unbedded pumice-lapilli tuff and probably unwelded, but at the base the tuff is distinctly bedded with vitric as well as pumice-lapilli tuffs. The massive tuff is overlain by a nodular tuff and then come obviously eutaxitic welded tuffs. Much of the Capel Curig Tuff Group in this locality was therefore emplaced subaerially.

Examine the east–west-trending dolerite dyke cropping out in the depression 75 m northeast of the summit. This dolerite is one of a series with this trend which cuts the Cwm Tryfan Anticline and which probably infills tensional fractures produced during the folding. Notice that the thick wedge of Glyder Fach Breccias dies out and fails to reach this southeastern limb of the Cwm Tryfan Anticline.

Continue around the top of the cliffs of Gallt yr Ogof. The Capel Curig rhyolitic tuffs, bedded lapilli tuffs, vitric and crystal-vitric tuffs are overlain by about 50 m sandstones dipping 50° southeast. The sandstones pass up into siltstones and slates, with a prominent dolerite sill near the base.

Descend to the exposures at Bwlch Goleuni and note that the southeasterly dip of the beds has decreased to 20° as a synclinal axial trace which separates the Cwm Tryfan Anticline from the Capel Curig Anticline is approached. Continue to Cefn y Capel and, at about [702582], slates with siltstones and fine-grained sandstones dip north at 10°, and the axial trace of the syncline has been crossed.

Walk north for about 1500 m to the old road parallel to the A5 and follow the track back to your transport.

EXCURSION 23. CAPEL CURIG TO ROMAN BRIDGE

This traverse covers the succession from the Capel Curig Volcanic Group, exposed in the Capel Curig Anticline, to the graptolitic slates overlying the Snowdon Volcanic Group exposed in the Dolwyddelan Syncline. The route takes in the summit of Moel Siabod and involves about 12 km of walking, some over rather rough ground. It may prove that many geologists would prefer to take two days over this excursion.

Use the IGS 1:25 000 Geological Special Sheet (Central Snowdonia) as an excursion map.
References: Francis and Howells (1973), Howells *et al* (1973), Williams and Bulman (1931) and Williams (1922).
1:50 000 OS sheet No. 115; or 1:25 000 sheets Nos SH65 and 75.

Part 1: Llynnau Mymbyr

Park in one or other of the large lay-bys on the A4086 at Llynnau Mymbyr. Pass through a gate at [706575] onto the hillside and walk north up the hill keeping to the east of a prominent wall.

The first exposures, near to the wall and about 150 m north of the road, are in the lowermost of the tuffs comprising the Capel Curig Volcanic Group. The formation has been called the 'Garth Tuff' by Francis and Howells (1973). The rock is a welded, rhyolitic tuff. There are no large lapilli, but even so the eutaxitic structure is obvious. A strongly nodular horizon is developed immediately above the eutaxitic horizon in a massive, crudely columnar-jointed tuff. The intensity of the nodular development decreases upward and nodules become sporadic as the massive tuff passes into a well bedded tuff and, eventually, into finely laminated tuffaceous sediments.

The Garth Tuff is succeeded by a dark blue-black silty slate which is poorly exposed, but then comes a crag of massive, coarse-grained sandstones showing large-scale cross bedding. These give way upwards to parallel-laminated and sinusoidally rippled, finer-grained sandstones which, in the upper third of the exposure, possess horizons rich in large brachiopods. The fossils are poorly preserved and specimens are very difficult to remove. The beds dip 30° northwest and lie in the northwestern limb of the periclinal Capel Curig Anticline.

This provides a good point from which to look southwest across Nant-y-gwryd to the southwestern closure of the periclinal Capel Curig Anticline, which is beautifully delineated by the curving strike of the sandstones in the hillside to the south and southwest of Garth.

The sandstones are followed by an impersistent black slate and then by a second tuff, the Racks Tuff. This is a massive tuff which is sporadically nodular and columnar-jointed, but apparently unwelded. The base of the tuff is remarkable. At the wall the black slate cuts completely through the tuff in what Francis and Howells (1973) described as a large 'flame-like' structure. In a traverse eastward from the wall, at 80 m the columnar jointing is no longer evident in the tuff and soft-sediment deformation structures are present. At 100 m the black slate reaches to within 3 m of the top of the tuff horizon and soft-sediment deformation structures are intense. This is repeated in a further 10 m, and at [709579] the tuff becomes discontinuous and occurs as pods separated by pelite.

The Racks Tuff is overlain by thin slates and silty slates and then by the Dyffryn Mymbyr Tuff. This is about 8 m thick and consists of strongly cleaved, bedded rhyolitic tuffs and tuffaceous sediments passing up into siltstones and slates.

Francis and Howells (1973) argued that, despite the welded character of a portion of the Garth Tuff in this traverse and a portion of the Racks Tuff in other localities, the evidence suggests that each of the tuffs accumulated in water. Certainly the Racks Tuff has been affected by contemporaneous deformation of a soft-sediment type; the 1:25 000 map and the evidence of Francis and Howells (1973) illustrate large structures (up to 250 m across) resembling load casts on the base of the Garth Tuff. There can be no doubt that the Garth Tuff and the Racks Tuff accumulated extremely rapidly on soft, water-saturated sediments, and it may be that, as Francis and Howells suggest, the volcanic rocks represent accumulation from sub-marine ash-flows.

Part 2: Bryn-engan to Moel Siabod Summit

Return to the road and take the footpath across the Nant-y-gwryd at the outlet from Llynnau Mymbyr.

Follow the footpath across the Garth Tuff up through the plantation to the higher roadway. Turn west and walk along the roadway cutting to [714573]. The outcrop is in the top of the Garth Tuff. On the weathered surface at the eastern end of the cutting, crudely bedded tuff passes up into well bedded tuff consisting of accretionary lapilli. These lapilli lie on a planar erosion surface which cuts deformed (soft-sediment type) bedded tuff. Some 60 cm above this horizon an erosion surface is cut into the beds of accretionary lapilli, and the bedded tuff lying on this surface shows evidence of vertical accumulation. The highest horizon exposed shows trough cross beds 60 m across, and the foresets have been modified by soft-sediment deformation. The evidence suggests accumulation of this upper part of the Garth Tuff in shallow water under the influence of intermittent strong currents.

Return east to the stream and follow the footpath up the hill for 400 m. Turn west and cross the stream to prominent exposures of the Racks Tuff which here is a cleaved, rhyolitic, pumice tuff. The tuff boundaries can be seen to be displaced slightly by minor east–west dip faults. The tuff is nodular near the top and has a eutaxitic, welded interior. The sandstones overlying the tuff are sometimes parallel-laminated, but others show symmetrical and asymmetrical ripples, washouts and small scours. The bedding dips 32° east and cleavage dips 75° northwest, so that bedding/cleavage intersections plunge 25° towards 040°. The exposures therefore lie in the southeastern limb of the Capel Curig Anticline, not far from the northeastern closure.

Traverse east across underlying cross bedded sandstones to the Garth Tuff. The top is well bedded pumice tuff with small-scale cross bedding and a couple of interbedded, thin vitric tuffs. A coarse nodular horizon, with nodules from golf ball to cricket ball size, is developed beneath the bedded top, and below this nodular zone is a massive, presumably welded, zone. Pumice lapilli are rare and a eutaxitic structure is not easy to see.

Follow the base along the strike. It is immediately apparent that the base is irregular in the manner mapped and described by Francis and Howells (1973). The basal facies is usually well bedded and sometimes laminated. Depressions in the underlying sediments are filled with an assortment of coarse lapilli tuffs, fine vitric tuffs and all possible admixtures of these two. The downward protruberances of tuff into the underlying sediments have been interpreted by Francis and Howells as down-sags analogous to load casts. Rapid emplacement of the tuff as a sub-marine ash-flow was thought to have caused liquefaction and yielding of the sediments.

Part 3: Moel Siabod to Clogwyn Mawr y Ffridd

Set off southward for the summit of Moel Siabod across the Garth Tuff and the Racks Tuff, across the sandstones, and so on to slates and siltstones. The beds dip at about 25° southeast in this limb of the anticline and the slates contain occasional horizons with concentrations of shelly fossils. The summit ridge of Moel Siabod consists of a thick dolerite sill. The rock is coarse-grained and gabbroic in many places, with augite and plagioclases over 1 cm across. Coarse columnar jointing is common.

Descend southwest from the summit for 250 m, and then turn southeast down a broad shoulder across the dolerite and onto slates and siltstones dipping southeast at about 25°. Cleavage now dips northwest at

60°. At [707541], a thin rhyolitic tuff dipping southeast occurs overlying slates and is, in turn, overlain by a thin tuffaceous sandstone. Continue southeast down the shoulder, across a dolerite sill, onto a second tuff overlain by a thicker sandstone unit, and then onto slates. At [707534] exposures of bedded rhyolitic tuffs representing the Lower Rhyolitic Tuff Group include crystal-vitric tuffs, lapilli tuffs and tuffaceous sediments. The tuffs are preserved in a syncline plunging at about 15° towards 070°.

Continue southeast across northerly dipping sandstones and two dolerite sills onto slates and siltstones. The slates become horizontal at [707527], cleavage dips northwest at 55°, and an anticlinal axial trace trends at 070° through this locality because south of this point beds dip southeast. Continue south onto a thick dolerite sill, the top and bottom of which are cleaved and amygdaloidal, whereas the interior is homogeneous and coarse-grained. The sill is overlain by a tuffaceous slate in which cleavage strikes at 056° and dips at 35° northwest. The measurements indicate that the cleavage becomes less steeply inclined as the traverse passes south from the Capel Curig Anticline.

Continue towards Clogwyn Mawr y Ffridd across a 150 m wide exposure gap. The prominent ridge of Clogwyn Mawr y Ffridd is formed by outcrops of the Lower Rhyolitic Tuff Group. In a traverse across the strike to Ffridd the first beds met with on the north of the feature are cleaved, rhyolitic, crystal-vitric tuffs in which bedding dips 68° towards 332°, whereas cleavage dips 45° towards 353°. The beds are therefore overturned to the southeast, and the exposure is in the overturned northwestern limb of the Dolwyddelan Syncline. This relationship between bedding and slaty cleavage is maintained throughout Clogwyn Mawr y Ffridd. About 80 m of bedded tuffs are exposed in the traverse and include vitric tuffs, lapilli tuffs and tuffaceous sediments as well as crystal-vitric tuffs.

Part 4: Clogwyn Mawr y Ffridd to Roman Bridge

From Ffridd walk west along the road for 800 m and, just before the bridge over Ceunant Ty'n-y-Ddol, turn north through a gate to exposures in the Bedded Pyroclastic Group at [699517]. The exposures are of bedded crystal-lithic and lapilli tuffs of basic composition. Lapilli and occasional small blocks are often of vesicular basalt but occasionally of rhyolite. The same bedding/cleavage relationship holds as at Clogwyn Mawr y Ffridd and therefore the exposures are in the overturned northwestern limb of the Dolwyddelan Syncline.

Continue along the road to Coed Mawr along the strike of the bedded basic tuffs. Take the footpath south from Coed Mawr towards the stream, crossing sporadic exposures of the Upper Rhyolitic Tuff Group. The outcrops are of cleaved, crudely bedded, rhyolitic crystal-vitric tuffs of the Upper Rhyolitic Tuff Group. Bedding/cleavage relationships indicate that the traverse is still within the overturned northwestern limb of the syncline.

Walk downstream to the bridge and so to the old Hendre Quarry [697512], which is cut in black pyritous slates overlying the Upper Rhyolitic Tuff Group and is sited at the southwestern closure of the Dolwyddelan Syncline. As a consequence, bedding dips at 30° northeast, whereas cleavage dips at 45° northwest and the slates cannot be parted along the bedding. Take the road back to Blaenau Dolwyddelan and then the footpath southeast to the old quarry at [700515]. The quarry itself is water-filled and great care is needed if you decide to examine the slates *in situ*. Bedding and cleavage are practically coincident and therefore graptolites are readily obtained from the waste tips. The fauna indicates the presence of the zone of *D. clingani*. The sooty black slates sometimes show a fine parallel lamination and were obviously deposited under tranquil marine conditions beneath the reach of wave base.

Take the footpath east to the exposures at the barn by the bridge over the railway at [705515]. These outcrops of the Upper Rhyolitic Tuff Group lie in the right-way-up southeastern limb of the Dolwyddelan Syncline, as can be readily determined from the bedding/cleavage relationships, since now the bedding dips 40° northwest, whereas the cleavage dips at 55° in the same direction. The rocks are dark blue, crudely bedded, cleaved, crystal-vitric tuffs of rhyolitic composition.

Take the track eastward alongside the railway to exposures east of Gorddinan. The rocks are bedded crystal-lithic and lapilli tuffs of rhyolitic and intermediate composition of the Lower Rhyolitic Tuff Group. Further exposures of the Lower Rhyolitic Tuff can be examined in the road cutting before Roman Bridge Station. This rock is a white-weathering, coarse-grained feldspathic tuff. The tuff is faulted against underlying slates and sandstones and the fault can be seen in a gully immediately east of the cutting.

Continue southeast along the road to the main A470 road. Examine well bedded slate and siltstone exposures on the way. Walk southwest around the double bend to roadside exposures on the eastern side of the road at [711510], where cleaved sandstones yield abundant brachiopods and a few gastropods. Bedding dips 40° northwest, whereas cleavage dips 83° northwest, and the traverse ends here in the sediments beneath the volcanics in the right-way-up southeastern limb of the Dolwyddelan Syncline.

EXCURSION 24. THE A5 SECTION: BETHESDA TO OGWEN COTTAGE

This low-level excursion is provided for those who are unable to walk the mountains. It can be a useful two days' work if the weather turns particularly bad.

The purpose is to examine a section from the Lower Cambrian into the Caradoc, in exposures from the Arfon Anticline to the Cwm Idwal Syncline. The rock sequences to be seen are tabulated in figure 4 (column 2) and in figure 5 (columns 6 and 7).

Use the excursion geological map, figure 48 (route B), and the IGS 1:25 000 Geological Special Sheet (Central Snowdonia).
Reference: Williams (1930).
1:50 000 OS sheet No. 115; or 1:25 000 sheets Nos SH65 and 66.

Locality 1: Ogwen Bank

A car can be left in the lay-by on the A5 at Ogwen Bank [627654].

Walk north along the main road towards Bethesda to outcrops on the eastern side of the road in purple slates with thin silty beds dipping southeast. These belong to the Purple Slate Formation of the Lower Cambrian Slates. Return to the lay-by and take the footpath leading northeast up the hillside. The path runs along the junction of Purple Slates to the northwest and overlying Green Slates to the southeast. The purple and green slates accumulated beneath the reach of wave base under quiet conditions. The occasional thin siltstones and rare fine-grained sandstones are probably the result of sporadic, far-travelled turbidity currents. It is from the horizon of the green slates that a sparse Protolenid fauna of late Lower Cambrian age was obtained.

After about 300 m turn southeast across the Green Slates onto the crags formed by good exposures of the Bronllwyd Grits. Evans (1968) claimed that there is passage at this locality from the Green Slates into the overlying Bronllwyd Grit. The lowest exposures are 12 m of coarsely conglomeratic greywackes in which the clasts are of rhyolitic volcanics and mud pellets. The beds dip eastsoutheast at about 60° and lie in the southeastern limb of the Arfon Anticline. They are succeeded by about 50 m of coarse greywackes with individual beds 0·5–1 m thick. Horizons rich in mud pellets still occur. The highest beds in the formation are greywackes, commonly about 30 cm thick, with thin interbeds of mudstone. The sandstones usually show a graded interval, followed by a parallel-laminated interval. Load casts, groove casts and flute casts are present. The rocks have the characters of proximal turbidites and represent a sudden increase in turbidity current activity.

Continue southeast across the Bronllwyd Grit onto grey-black slates, silty slates and thin, fine-grained sandstones and occasional conglomeratic sandstones of the Maentwrog Beds. Evans (1968) claimed that there is passage from the Bronllwyd Grit into the Maentwrog Beds at this locality, and suggested that the Bronllwyd Grit is Middle Cambrian. Most workers, however, interpret the base of the Maentwrog Beds as a disconformity and regard the Bronllwyd Grit as topmost Lower Cambrian. This is the interpretation illustrated in figure 4.

Cross to prominent crags at [633653] near a footpath that obliquely ascends the valley side. At this locality, a dyke-like dolerite intrusion cuts the grey-black slates. Continue up the path and pass through the gate to exposures of dark, grey-black, laminated, silty slates with thin sandstones. One of the fine-grained, white-weathering sandstones is 60 cm in thickness. Towards the top the Maentwrog Beds become distinctly flaggy with many siltstones and fine-grained sandstones which weather white. A wealth of sedimentary structures can be seen including washouts, cross lamination, convolute lamination, load casts, flute casts and flame structures. The siltstones and fine-grained sandstones are probably distal turbidites and the dark slates and silty slates represent the background sedimentation. Accumulation probably occurred beneath the reach of wave base.

The Maentwrog Beds are succeeded by crags of grey-weathering granophyre. The rock consists of a fine-grained matrix of quartz and feldspar with specks of chlorite after biotite; set in the matrix are sparse feldspar phenocrysts up to 5 mm across. The granophyre is emplaced as a sill-like mass within Ffestiniog quartzites which can be examined in the corner of the field below the granophyre crag. The quartzites are better exposed about 300 m northwest of the crag where burrowed, massive, yellow sandstones, conglomerates, flaggy quartzites and thin slates crop out. Cross bedding is common and, apart from abundant *Skolithos*, *Cruziana* can be found. The sediments probably accumulated close to low water mark.

The Ffestiniog Beds are overlain by soft, blue-black slates of Llanvirn lithology, which were mapped as such by Evans (1968). Exposures are plentiful on the hillside. The absence of sedimentary structures, apart from parallel lamination, suggests accumulation below wave base.

Figure 52. View from Ty'n-y-maes on the A5 looking southwest to Carnedd y Filiast, showing the distribution of the main rock groups. (Photograph by M S Hobbs.)

Locality 2: Tai Newyddion

Return to the A5 and take the old road over Pont y Ceunant.

Look to Fronllwyd and see the craggy Bronllwyd Grits generally dipping southeast beneath the softer Maentwrog Beds. The overlying Ffestiniog Beds form Carnedd y Filiast and the big bedding planes can be seen dipping at about 50° southeast. The resistant Ffestiniog Beds strike down the hillside towards Tai Newyddion (figure 52). Make for the low exposures at [626637] about 400 m due east of the T-junction at [630637]. The hillside here consists of strata near the top of the Maentwrog Beds. Thin greywacke sandstones are interbedded with slates. Sedimentary structures to be seen include cross lamination, convolute lamination, flute casts, load casts and flame structures.

Contour around to the prominent crags immediately west of and above Tai Newyddion. Quartzites and granule conglomerates crop out. Bedding surfaces are commonly rippled and undersurfaces often show biogenic sedimentary structures, especially trails such as *Cruziana semiplicata*.

Cross the wall to the south and keep just above the lower of the two walls running south along the valley side. Further cross bedded and ripple-marked quartzites and conglomerates are exposed. The wall then turns and descends to the valley floor and at the bend the quartzites are overlain by tough, grey silty slates mapped by Williams (1930) as Llanvirn, but which are possibly of Ffestiniog age. Cross the exposure gap to crags of soft blue-black slate of Llanvirn lithology. The slope now becomes steep and it is best to descend to the old road.

Locality 3: Cwm Bual

Follow the old road southeast past Maescaradog, for another 400 m.

At the gate a path leads due north to a prominent *roche moutonée* projecting from the alluvial floor of Nant Ffrancon at [638624]. It is formed by a cleaved, buff-weathering quartz–dolerite dyke which can be seen trending southwest up Cwm Bual where, along the north side of the stream, it is seen to be stepped in an *en echelon* manner. (This is not the dyke drawn on Williams' (1930) map as cropping out in Cwm Perfedd.) The dyke can also be conveniently examined in the rough pasture immediately west of the road (see figure 53).

South of the gate across the road, the stream from Cwm Bual has built a low fan. Examine the soft, blue-black slates of Llanvirn lithology at the apex of the fan. The soft slates persist to the stream running down the hillside 100 m south of the barn at [638619]. Just beyond the stream tough, splintery, grey-green slates, siltstones and fine-grained sandstones crop out, which occupy an anticlinal structure plunging at about 70° to 210° and are probably representatives of the Ffestiniog Beds. The beds have been mapped as Llanvirn by Williams (1930).

Locality 4: Blaen-y-nant

Continue along the road beyond Pentre where a fan of debris 450 m around the base, has been derived from Cwm Coch.

The fan conceals slates of Llanvirn lithology and the crags beyond are formed by the Bwlch y Cywion microgranite. At the gate across the road 300 m north of Blaen-y-nant the contact of the Bwlch y Cywion intrusion with Llanvirn slates is exposed. The junction dips southeast and the slates at the contact are baked, but the cleavage has not been annealed. The marginal facies of the intrusion is a flinty rhyolite but it rapidly becomes a felsite and then a microgranite inward from the junction, which is irregular in detail.

Continue up the road and, just after the turn off to Blaen-y-nant, the slates at the contact are probably now Caradocian Glanrafon slates and have been contact-metamorphosed to spotted slates with porphyroblasts of cordierite. The traverse is now on ground covered by the IGS 1:25 000 map. Some 50 m further up the road the Glanrafon slates show a slaty cleavage, S_1, refolded about low-angled axial surfaces. The sideways-facing folds that are developed represent F_2 structures and they are accompanied by a strong, axial-planar crenulation cleavage, S_2. Both F_2 and S_2 are particularly well developed in slates and siltstones further on at the bridge over the stream a little way up the road at [642607].

Some 50 m further on, beyond the bridge, the Glanrafon slates possess very prominent sub-horizontal kink bands, which may be F_2 structures. A cleaved dolerite can be seen striking away southwest up the hillside. The margins are fine-grained, but grain sizes increase to gabbroic inwards.

The road then crosses pale-weathering, cleaved, rhyolitic tuffs, regarded originally by Williams (1930) as rhyolite and keratophyre lavas. Kink

Figure 53. View from a point of the A5, 500 m southeast of Ty-gwyn looking across at Foel Goch and Mynydd Perfedd, and showing the distribution of the main rock groups. (Photograph by M S Hobbs.)

bands are prominent in the roadside exposures. The road now enters a cutting where, on the southwestern side, a deeply spheroidally weathered, Tertiary dolerite dyke crops out. The fresher rock inside the spheroids is a green, vesicular, porphyritic, olivine dolerite. The dyke has the usual northwesterly trend.

The tuffs are overlain by further silty Glanrafon slates now well beyond the contact aureole of the Bwlch y Cywion intrusion and, 60 m before the climbing hut, an amygdaloidal dolerite sill cuts across the road. The sill is indicated on both the IGS map and on the map of Williams (1930) as terminating on the hillside before the road is reached. At the climbing hut the slates are overlain by green, tuffaceous sandstones and siltstones with bands of porcellanous, rhyolitic vitric tuff. The sandstones become coarser-grained upward and the tuffaceous sandstones contain conglomeratic horizons in which the clasts are of well rounded rhyolite.

The road now bends sharply eastward and the sandstones are succeeded here by a cleaved and crudely bedded agglomeratic pumice-lapilli tuff of rhyolitic composition. The tuff—a water-lain, bedded facies of the Capel Curig Volcanic Group—is overlain by massive sandstones, the Gwastadnant Grits which, near the conifer plantation, are strongly cross bedded. The grits are rich in plagioclase and lithic fragments of andesitic composition so that they are essentially volcanic sands.

The Glanrafon succession began with slates in which occasional thin siltstone and fine-grained sandstone laminae occur. Parallel lamination is usual, but the siltstones and fine-grained sandstones often show cross lamination. Tuffs and siltstones become important towards the top of the slaty succession, and finally pass into the bedded tuffs and often conglomeratic volcanic sands of the Gwastadnant Grits. The sediments record a gradual shoaling and an increase in the amount and grain size of volcanic detritus.

Locality 5: Idwal Cottage

10 m inside the plantation the Pitt's Head ignimbrite succeeds the Gwastadnant Grit. The basal facies of the ignimbrite is a cleaved, non-welded, rhyolitic, pumice-lapilli tuff which, behind the former chapel (now a YHA annexe) passes up into a strongly eutaxitic welded tuff. Further sandstones, often tuffaceous, with interbedded siltstones and slates follow and a synclinal axial trace within the Cwm Idwal Syncline is crossed at Idwal Cottage. The closure of the fold at Ogwen is not simple: one axial trace trends from Idwal Cottage northnortheast across through Pont Pen-y-benglog and up the opposite hillside towards Pen-yr-Ole Wen. The main axial trace, however, runs northeast across Llyn Idwal and this structure closes at Llyn Ogwen.

The progressive shoaling evident during the accumulation of the Glanrafon slates and sandstones culminated with emergence and the subaerial accumulation of the Pitt's Head ignimbrite.

Return down the old road to the transport.

EXCURSION 25. THE A5 SECTION: OGWEN COTTAGE TO GALLT YR OGOF

This low-level excursion is a continuation of the previous one; the references are therefore the same. The aim is to examine the Caradoc succession as developed in the Cwm Idwal Syncline and the Cwm Tryfan Anticline.

Use the IGS 1:25 000 Geological Special Sheet (Central Snowdonia) as an excursion map.

Part 1: The Northern Shore of Llyn Ogwen

The excursion begins in the northwestern limb of the Cwm Idwal Syncline at the telephone kiosk on the A5 at Pont Pen-y-benglog, at the outlet from Llyn Ogwen [649604].

A eutaxitic facies of the Pitt's Head ignimbrite is exposed in the roadside cutting. The eutaxitic, rhyolitic, pumice-lapilli tuff is overlain to the southeast by a cleaved and bedded facies dipping 52° southeast.

Take the footpath which leads along the northern shore of Llyn Ogwen. A synclinal axial trace is crossed, which can be seen outlined in the bedded tuffs and tuffaceous sediments above the ignimbrite, and the beds now dip northwest. It is underlain by green, tuffaceous sandstones. A minor fold-pair plunging southwest is crossed in the tuffaceous sandstones and then, just over 300 m from the telephone kiosk, the sandstones are cut by a sill-like acid intrusion. The intrusion is columnar-jointed with a flow-banded, rhyolitic margin which passes rapidly inward to a more homogeneous granophyric micrognanite interior. At the short, old wall leading down to the lake at about [652605], an east–west-trending dyke of dolerite cuts the granophyre. This is a representative of the east–west-trending swarm that fills tensional fissures in the Cwm Tryfan–Cwm Idwal fold-pair and the dyke can be seen to be virtually uncleaved and was presumably emplaced at a late stage of the F_1 fold phase.

Trudge along exposures of the microgranophyre, with its good columnar jointing and occasional flow-banding and flowage-folding, to impressive exposures of parallel-laminated and cross bedded massive sandstones. The sandstones, which are about 50 m thick, contain large cubes of pyrite up to 3 cm across, and the sandstones dip northwesterly at about 55°.

Cross the sandstones onto underlying rhyolitic tuffs of the Capel Curig Volcanic Group. The highest beds are beautiful bird's-eye tuffs in which the accretionary lapilli have been flattened in the plane of the cleavage to ellipsoids. On the assumption that there was no viscosity contrast between the accretionary lapilli and the enclosing matrix, strain measurements indicate a tectonic thickening in these beds of approximately $\times\frac{5}{3}$. The bird's-eye tuffs are underlain by bedded lapilli tuffs and by massive tuffs, some of which may be ignimbrites.

Towards the head of the lake the ground is largely drift-covered, so continue around the head of the lake to the A5 via Pont Tal-y-llyn.

Part 2: Pont Tal-y-llyn

Climb over one of the several stiles onto the rough ground south of the A5 at the head of Llyn Ogwen.

The exposures at about [665603] are in a massively columnar-jointed, flow-banded and flowage-folded rhyolite. This is the rhyolite sill that forms that part of the eastern face of Tryfan above the Heather Terrace.

Traverse east across the rhyolite onto a thin band of sandstones, siltstones and slates about 10 m thick. The beds occupy a largely grass-covered rake and dip northwest at 53° into the Cwm Idwal Syncline. The sediments are immediately underlain by bedded pumice-lapilli tuffs of rhyolitic composition. The tuffs rapidly become massive downward, fine-grained and devoid of pumice lapilli, although rare blocks of rhyolite can be found. The tuff is strongly cleaved. It is apparent that the total thickness of the rhyolitic tuffs of the Capel Curig Volcanic Group is considerably greater in this eastern limb of the Cwm Idwal Syncline than it is in the western limb. Furthermore, ignimbrites are included in the group and therefore a significant proportion of the tuffs were emplaced subaerially.

The tuffs are underlain by parallel-laminated sandstones, tuffaceous sandstones, siltstones and slate. The beds pass down into sandstones which are highly tuffaceous and in which crystals and pumice lapilli tend to weather out. These largely volcanic sands are often well graded, and trough cross bedding and washouts are common. Some of the cross bedded and cross laminated sandstones possess individual foreset laminae which are very rich in heavy minerals. The beds clearly accumulated in shallow water under the influence of strong currents.

Part 3: The Axial Region of the Cwm Tryfan Anticline

Cross some exposures in which bedding planes of the sandstone are very prominent onto a massive, unbedded rhyolitic pumice-lapilli tuff which strikes down the hillside to the farm Wern [673604]. The tuff is crudely columnar-jointed but does not appear to be welded. It is probably the lateral equivalent of the rhyolitic tuff seen in Excursion 24 at Blaen-y-nant.

Cross over the stream flowing north from Cwm Tryfan over silty slates. The slates are underlain by a further rhyolitic, pumice-lapilli tuff and at [676599] a rounded anticlinal hinge is exposed in a low east–west-trending cliff.

Near the stream, in the western limb of this minor fold, the base of the tuff can be seen to be discordant on the small scale, and rests on an irregular erosion surface cut in underlying sandstones. Continue around the closure of the fold along the base of the tuff at the foot of the low cliff to the low spur which runs northeast towards the A5. Unmapped flow-banded and flowage-folded intrusive rhyolites occur amongst the lapilli tuff here at [677601]. A synclinal axial trace runs along the spur and the fold plunges northeast less steeply than the slope. The folds represent minor folds within the major Cwm Tryfan Anticline (see figure 51).

Continue east across tuffaceous sandstones just above the old fence until the tuff is crossed once again, striking diagonally down the hillside. A thin horizon of slates is followed by sandstones cropping out on a craggy spur. The sandstones are parallel-laminated, and patchy cementation has resulted in weathering producing a series of ellipsoidal depressions up to 15 cm long which are concentrated along certain horizons. They dip at 45° southeast and the traverse now lies in the southeastern limb of the Cwm Tryfan Anticline.

The sandstones are overlain by the younger of the rhyolitic tuff horizons which here are well bedded, crystal-vitric and crystal-lithic tuffs. Contour across the slope over slates and siltstones to sandstones cropping out beyond Nant-yr-ogof. The massive sandstones are sometimes parallel-laminated, cross bedded or trough cross bedded. Towards the top they become very coarse-grained and conglomeratic with many rhyolite clasts. Many bedding planes are ripple-marked. Clearly, the suite of volcanic sands, silts, slates and tuffs exposed in the core of the Cwm Tryfan Anticline and lying stratigraphically beneath the Capel Curig

Volcanic Group accumulated under shallow marine conditions, perhaps often near low water mark.

Part 4: Gallt yr Ogof

At the spur running northeast from Gallt yr Ogof, begin to climb up over the hillside onto the Capel Curig Volcanic Group once more. The sequence begins with strongly cleaved, bedded lapilli tuffs of rhyolitic composition. These are overlain by beds of vitric and crystal-vitric tuff, and then by massive, unbedded pumice tuffs which may be welded. These possible ignimbrites are succeeded by bedded vitric tuffs, and then by a 5–10 m wide, largely grass-covered hollow in which tuffaceous siltstones and slates crop out sporadically. The sediments are overlain by further vitric tuffs and bedded pumice-lapilli tuffs, and then comes a massive tuff in which nodules the size of tennis balls have developed. This nodular horizon is immediately overlain by a strongly eutaxitic welded tuff and it may be that the massive tuff, the nodular tuff and the eutaxitic tuff are part of a single ignimbrite sheet. Several distinct ignimbrite flows follow, each having a non-welded base which rapidly becomes welded upward.

At [692594] a thick lens of slates and sandstones occurs within the volcanic unit. Brachiopods are common in the sediments, and good internal and external moulds can be picked up in the scree. The fossiliferous sediments are overlain by cross bedded crystal-lithic and crystal-vitric tuffs. Descend with caution over the steep slope cut in the tuffs into the cwm of Gallt yr Ogof and return to Llyn Ogwen via the old road.

EXCURSION 26. THE CARNEDDAU

This excursion is entirely within Caradocian sediments and igneous rocks, and provides a magnificent, though strenuous, day's hill walking. The succession is tabulated in figure 5 (column 7).

Use the excursion geological map, figure 54.
References: Diggens and Romano (1968) and Evans (1968).
1:50 000 OS sheet No. 115; or 1:25 000 sheets Nos SH66 and 76.

Part 1: Pont Tal-y-llyn to Cwm Lloer

Leave your car in one of the parking places at the head of Llyn Ogwen.

Cross the Afon Denau at Pont Tal-y-llyn and examine the outcrops by the track at [668606]. The rocks are rhyolitic tuffs of the Capel Curig Volcanic Group and here they are eutaxitic, welded, pumice-lapilli tuff dipping west at about 35° on the common limb of the Cwm Tryfan Anticline and the Cwm Idwal Syncline.

Take the footpath uphill parallel to the Afon Lloer. About 100 m above the footbridge bedded rhyolitic tuffs are exposed at the stream, whereas to the west are small exposures of flow-banded and flowage-folded rhyolites which are a continuation of the rhyolite sill of Tryfan seen across the valley to the south. The bedded tuffs in the stream are underlain to the east by the welded-tuff horizon seen earlier by Pont Tal-y-llyn. At this locality the welded tuffs have a nodular top and the horizon forms the ridge of well exposed ground east of the wall running parallel to and east of the Afon Lloer.

About 12 m further upstream at a bend to the northeast, at [668612], a dolerite sheet is exposed trending northeast–southwest.

Incidentally, there is a magnificent view into the Cwm Idwal Syncline from here, and the view of the major fold-pair of the Cwm Idwal Syncline and the Cwm Tryfan Anticline gets better as the hillside is climbed (see figures 50 and 51).

Continue uphill, parallel to the stream and through a prominent wall running parallel to the contours. On the eastern side of the stream at [667618], bedded rhyolitic tuffs have swung to dip at 30° to the west-southwest. As the stream swings to the west, so does the strike of the bedded tuffs, and at [665619] it is clear that a synclinal hinge has been crossed because the beds now dip at 35° to the southsoutheast. Cleavage

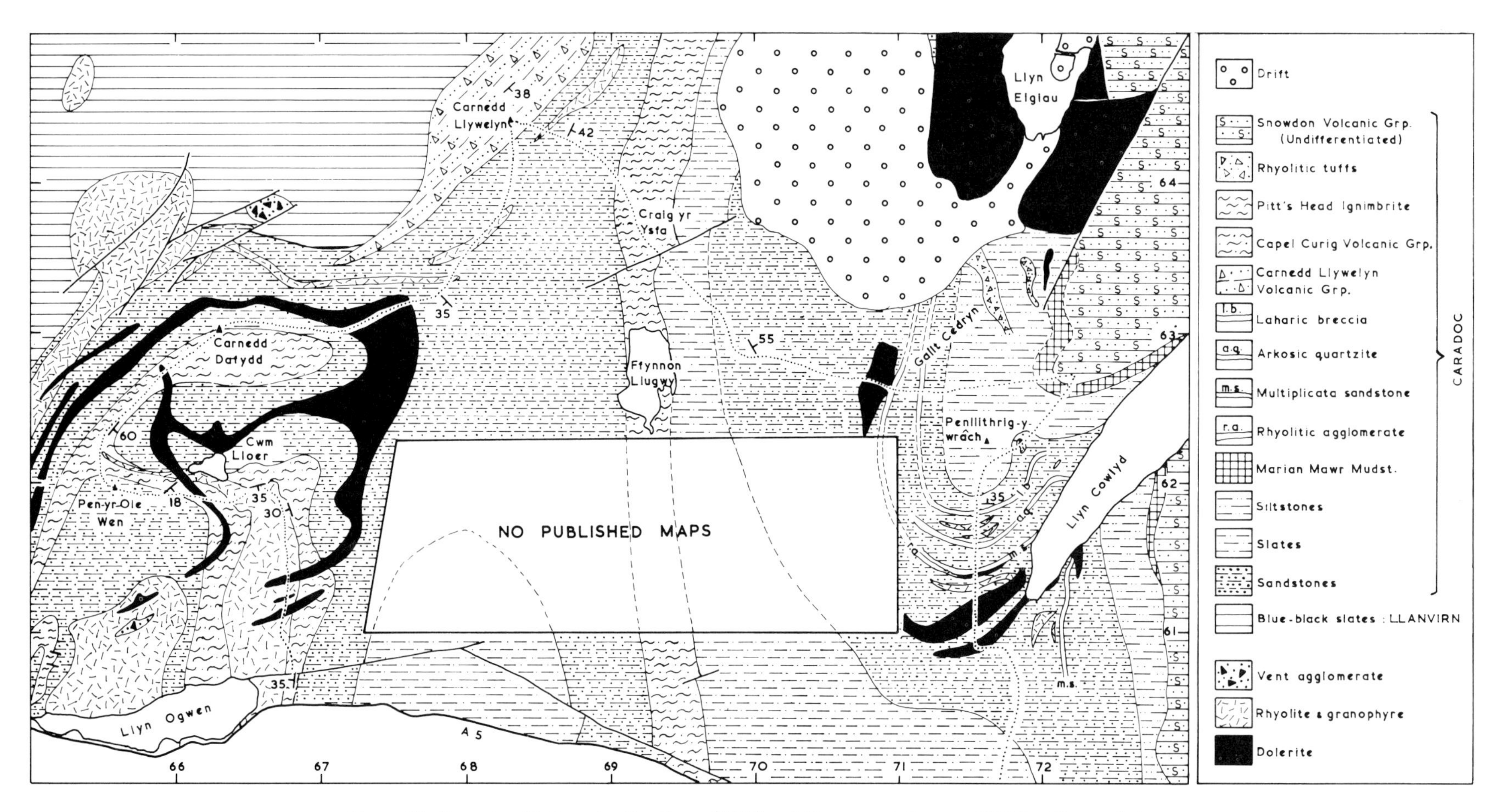

Figure 54. Excursion 26. Largely after Evans (1968) and Diggens and Romano (1968), with permission.

in the tuffs is vertical and strikes at 040°; the meagre structural data gathered thus far suggest a syncline plunging southwest. The excursion map shows a very complicated outcrop pattern but, happily, the geology is much simpler than the pattern suggests. In the first place, the axial trace is deflected by the presence, at Llyn Ogwen in the eastern limb of the fold, of a thick granophyric intrusion; and in the second place the great hollow of Cwm Lloer has taken a large bite out of the northwestern limb, leaving beds exposed on each side of the ridge forming the southern wall of Cwm Lloer. The slopes of Carnedd Dafydd, immediately to the south of the summit, are formed by a sheet of Capel Curig Volcanic Group rocks dipping southeast less steeply than the slope of the ground.

Part 2: Cwm Lloer to Pen-yr-Ole Wen

Follow the path westward up the ridge leading to Pen-yr-Ole Wen.

The flow-banded rhyolite sill is overlain by bedded rhyolitic tuffs, including tuffs rich in accretionary lapilli (bird's-eye tuffs). A thick dolerite sheet with a knobbly weathered surface is crossed, and then the path crosses a eutaxitic welded tuff before the bedded tuffs appear again, dipping at 18° to the southeast. The bedded tuffs, the youngest members of the Capel Curig Volcanic Group, are overlain by sandstones and siltstones of the Upper Glanrafon Beds which are well exposed over the last 200 m leading to the summit. Look south to the Cwm Tryfan Anticline and the Cwm Idwal Syncline (figure 55).

Part 3: Pen-yr-Ole Wen to Carnedd Dafydd

Take the path leading along the ridge towards Carnedd Dafydd.

The route takes you into the underlying Capel Curig Volcanic Group once again. At the col, turn east to the head of Cwm Lloer where fossiliferous sandstones are exposed beneath the volcanics. Internal and external moulds of orthids are readily collected. Continue along the ridge from the col over coarse-grained lithic tuffs and finer-grained crystal-vitric tuffs, all of which are of rhyolitic composition, and dip at 60° to the southeast. The traverse then crosses a nodular tuff in which the nodules are commonly the size of cricket balls, and then an agglomeratic tuff, which again is of rhyolitic composition.

Carry on past the large prehistoric cairn onto a cleaved, unbedded, pumice tuff and then on to a wall about 20 m long built on a very coarse-grained, pale-weathering, blocky jointed dolerite. Climb up the remaining 250 m across cleaved, rhyolitic, pumice-lapilli tuffs to the summit of Carnedd Dafydd.

Part 4: Carnedd Dafydd to Carnedd Llywelyn

Descend from the summit along the track towards Carnedd Llywelyn over the rhyolitic tuffs of the Capel Curig Volcanic Group onto underlying sandstones, siltstones and slates. The sandstones show large-scale cross bedding, and the siltstones cross lamination. Some bedding planes are ripple-marked. The beds contain a shelly fauna and good specimens of brachiopods are readily obtained. The sandstones are underlain by a very coarse-grained gabbroic sill which crosses the ridge. The mass is coarsely columnar-jointed and the weathered surfaces are knobbly. Plagioclase and augite crystals sometimes exceed 3 cm in diameter. The ridge narrows, and interbedded sandstones and siltstones underlie the sill and dip 35° southwest; the traverse now coincides with the axial trace of the major syncline. The fold dies out rapidly to the northeast, however, as the dips swing to the southeast over the next few hundred metres of the route.

About 300 m further northeast along the ridge from the base of the gabbroic sill, a thin sill of strongly cleaved but still recognisably flow-banded and flowage-folded rhyolite is emplaced within the sandstones. The underlying sandstones contain a plentiful shelly fauna and good specimens of orthids can be readily collected.

A second rhyolite sill crosses the narrow ridge towards its northeastern end. It is strongly autobrecciated and nodular, particularly near the margins, and it is underlain by tuffaceous sediments, slates and sandstones which persist almost to the summit of Carnedd Llywelyn.

The summit consists of an amygdaloidal basalt underlain to the northwest by a crudely bedded, cleaved basic agglomerate and basic tuffs. These basic volcanic rocks belong to the Carnedd Llywelyn Volcanic Group.

Part 5: Carnedd Llywelyn to Pen Llithrig-y-wrâch

Take the ridge leading southeast from the summit towards Craig yr Ysfa.

The basalt is overlain by basaltic and andesitic tuffs and tuffaceous sediments which comprise the topmost part of the Carnedd Llywelyn Volcanic Group, and then by a thin, flow-banded and autobrecciated, nodular rhyolite sill. Coarse-grained, massive tuffaceous sandstones succeed the sill. The sandstones show parallel lamination and also, on occasion, cross bedding. There is no reason to assume that the Carnedd Llywelyn Volcanic Group is anything other than marine.

At [691641] the sandstones are overlain by a bedded, agglomeratic, rhyolitic tuff which is succeeded by a coarse-grained andesitic tuff containing small rhyolite clasts, and then by tuffaceous sandstones or re-worked tuffs. A very thick rhyolitic pumice-lapilli tuff then follows, which has a well developed eutaxitic structure and is strongly welded.

Figure 55. Panorama looking south from Pen-yr-Ole Wen.

This sequence of bedded tuffs and ignimbrites represents the Capel Curig Volcanic Group, and records shoaling as the sandstones were succeeded by coarse-grained, bedded tuffs and agglomeratic tuffs and, finally, emergence with the emplacement of the welded tuffs under subaerial conditions.

A northeasterly trending fault crosses the ridge and brings in interbedded sandstones and slates to form the ridge above Ffynnon Llugwy.

(This is a good point to break off the excursion if time is beginning to run out. If this should be the case, descend by the path to the west shore of Ffynon Llugwy. Follow the track to the new road at the outlet from the lake and follow the new road down to the A5.)

If sufficient time remains, continue southeast along the ridge to Pen-yr-Helgi-ddu across sandstones and siltstones of the Upper Glanrafon beds. At Pen-yr-Helgi-ddu the strike has swung round so that the sandstones

dip 55° towards 060°. Continue eastward down the broad ridge of Y Las-gallt and, at the bwlch at the foot of the climb up to Pen Llithrig-y-wrâch, examine the coarse-grained, columnar-jointed dolerite sheet. A thin agglomeratic rhyolitic tuff overlies the sill and is followed by siltstones and sandstones. A second agglomeratic tuff, again rhyolitic, crops out about 200 m beyond the dolerite, and is overlain by siltstones and fine-grained sandstones. These pass up into a series of well cleaved, laminated siltstones and minor slates which persist to the summit of Pen Llithrig-y-wrâch. Submergence followed the emplacement of the ignimbrites of the Capel Curig Volcanic Group. The overlying succession of sandstones passing up into siltstones and minor slates may indicate a progressive deepening of the sea. The sandstones are moderately well sorted and often cross bedded, however, and the siltstones are commonly cross laminated. Accumulation therefore probably occurred well within the reach of wave base under the influence of sometimes strong currents.

From the point at which the northwesterly projecting ridge of Pen Llithrig-y-wrâch is truncated by the cliffs of Gallt Cedryn, walk on a bearing of 085° for 900 m, until prominent exposures of rhyolitic tuffs are met with at the top of the slope above Llyn Cowlyd. These are the lowest members of the Crafnant Volcanic Group which here rest on largely unexposed mudstones, the Marian Mawr Mudstone. The lowest rocks are strongly cleaved, unwelded crystal-vitric tuffs, but these become welded upward and a eutaxitic texture is present. Pumice lapilli are rare, however, and consequently the welded character can only be proven by thin-section study. The abrupt change from mudstones to ignimbrites indicates very abrupt uplift above sea level.

Part 6: Pen Llithrig-y-wrâch to Capel Curig

Walk southwest towards a small crag at [718623] across initially unexposed ground, but then onto the banded Pen Llithrig-y-wrâch siltstones. The rocks are often cross laminated and occasional calcareous horizons yield a shelly fauna of Lower Longvillian age.

The small crag maps as a pod-like mass and has been interpreted as intrusive rhyolite. The outcrop shows a banded rock with bomb-shaped masses and lapilli-sized fragments of vesicular pumice in a matrix which appears to be clastic. Some of the bomb-shaped masses show sags into the banded material below and the bomb-like pieces are confined to certain horizons. The rock may well be an agglomeratic lapilli tuff rather than an intrusive rhyolite.

Continue southwest across the siltstones and then turn south down the southern spur of Pen Llithrig-y-wrâch onto sandstones dipping north at about 35°. At the 2000′ contour, the coarse-grained, grey-weathering sandstones contain a 15 m thick muddy horizon with bands rich in clasts of rhyolite up to 10 cm across. Diggens and Romano (1968) have interpreted this breccia as a lahar (that is, the product of a volcanic mudflow).

Descend over further coarse-grained sandstones to a thin sill of flow-banded and flowage-folded, white-weathering flinty rhyolite. The sill is underlain by grey-weathering, cleaved, fine-grained sandstones and siltstones, and then a strikingly white-weathering unit about 15 m in thickness is exposed. The lowest 3 m is bedded and strongly cleaved, but it passes upward into a cross bedded quartzite which yields a plentiful trilobite–brachiopod fauna of probable Lower Longvillian age towards the top.

The underlying siltstones are not well exposed on this line of the traverse so that the next 100 m is largely grass-covered, but then a very fossiliferous sandstone crops out and forms a feature 3 m high. This is the *Multiplicata* Sandstone, which is, in fact, about 12 m thick. It is a greyish green, tuffaceous sandstone, usually cross bedded, and contains horizons crowded with *Dinorthis multiplicata* (Bancroft).

The sandstone is immediately underlain by a coarse-grained dolerite sill about 12 m thick, and beneath which are rather poorly exposed parallel-laminated sandstones, and then a 10 m thick sill of white-weathering, flow-banded and flowage-folded rhyolite. The traverse then crosses poorly exposed sandstones underlain by another coarse-grained dolerite sill. Head for the junction of the leats across cleaved sandstones and siltstones, and finally another dolerite sill just before the leats.

Follow the track across unexposed ground southward towards the A5. As the slope steepens above the Afon Llugwy, Glanrafon slate, siltstones and fine-grained sandstones are again exposed. Gain the A5 about 1 km north of Capel Curig.

EXCURSION 27. CAPEL CURIG TO TREFRIW

The purpose is to examine the rocks of the Crafnant Volcanic Group together with the underlying and overlying strata. The succession is given in figure 5 (column 10).

A car can usually be left for the day at Capel Curig or in one of the lay-bys at Llynnau Mymbyr; but it is better to arrange a ‘pick-up’ at Trefriw if possible.

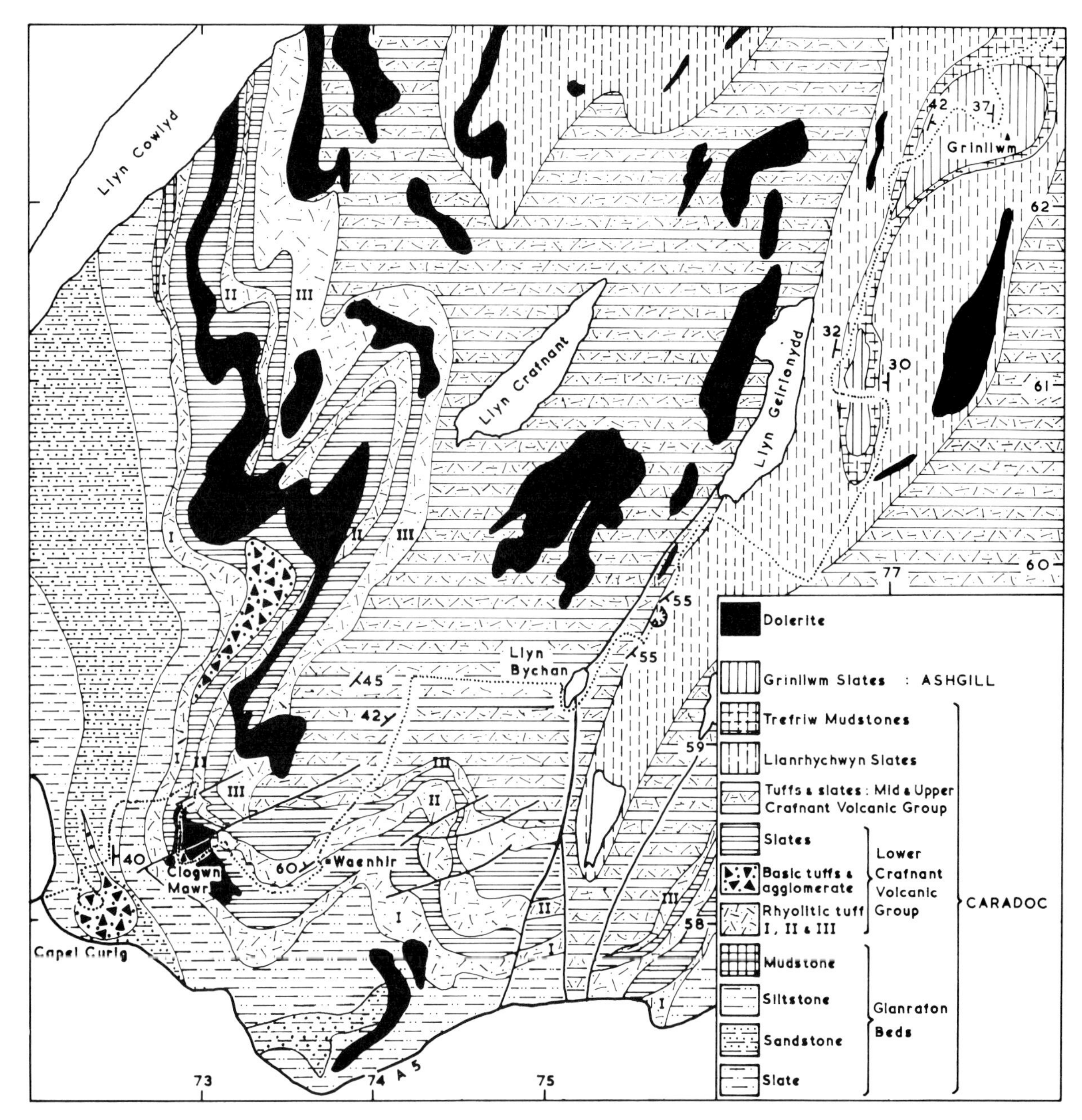

Figure 56. Excursion 27. Largely after Davies (1936), with permission, and partly after Howells *et al* (1973). The material from Howells *et al* is based on Geological Survey material and is reproduced by permission of the Director, Institute of Geological Sciences.

Use the excursion geological map, figure 56.
References: Davies (1936), Howells *et al* (1973) and Williams (1922).
1:50 000 OS sheet No. 115; or 1:25 000 sheets Nos SH75 and 76.

Part 1: Capel Curig to Waenhir

Take the footpath leading east from the A5 at a point just north of the main road junction at Capel Curig.

Examine the exposures on Capel Curig Hill, south of the footpath [724582]. The rocks exposed are agglomeratic, basaltic, pumice-lapilli tuffs. The outcrops suggest a pipe-like mass cutting the sandstone country rocks, and the mass is therefore interpreted as a vent agglomerate.

Immediately to the north of the vent and the footpath, sandstones are exposed striking at 345°. Interbedded with the easterly dipping sandstones is a 2 m thick, fine-grained sandstone with slump structures including overfolds, balled-up masses and oversteepened foreset beds. This disturbed horizon is overlain by bedded vitric tuffs with washouts several metres long, and the tuffs too show soft-sediment deformation.

Walk north along the strike for 250 m and then turn east and traverse uphill across the easterly dipping sandstones. Broken fossils (brachiopods) are common in the finer-grained sandstones and siltstones, and shell bands up to 10 cm thick are present. The bands tend to weather in because of their calcareous character. The brachiopods are usually poorly preserved.

Cross a 30 m exposure gap to a prominent crag at [727586]. The greyish white exposures are of the lowest member of the Lower Crafnant Volcanic Group, which is rhyolitic and about 55 m thick. The base is not seen, but the lowest rock exposed is a feldspar-rich tuff and about 1·5 m is exposed. It is overlain by 8 m of crystal-poor tuff with interbedded muddy horizons up to 3 cm thick. This appears to pass upward into an unbedded tuff with clasts of rhyolite and pumice lumps up to 10 cm across. All three facies are strongly cleaved. There is then an exposure gap of 15 m before massively jointed, grey-weathering, crystal-pumice tuffs which are crudely bedded at the base. These are succeeded by 8 m of poorly exposed, cleaved vitric tuff and then a thickly bedded vitric tuff which has undergone soft-sediment deformation. Some of the bands are rich in clasts of pumice and rhyolite, and are succeeded by an unbedded, columnar-jointed tuff which is nevertheless strongly cleaved and small pumice lapilli are flattened in the plane of cleavage.

Walk east across poorly exposed siltstones and slates to a prominent rhyolitic crag at [729587]. The crag is formed by the second tuff of the Lower Crafnant Volcanic Group and is underlain on the southwest and south by a strongly porphyritic dolerite with plagioclases up to 2 cm across in the marginal facies. Climb Clogwyn Mawr. Look northeast and see the tuffs of the Lower Crafnant Volcanic Group striking northeastward and dipping southeast. At Clogwyn Mawr the beds swing through the closure of a northeasterly plunging syncline to strike northeast and dip northwest. The Nant Geuallt runs along the axial trace of the syncline. Some 1500 m to the eastnortheast, the tuffs can be seen outlining a complementary anticlinal hinge (figure 57).

Descend from the dolerite of Clogwyn Mawr northeast to a knoll of rhyolitic tuff on the northwestern bank of the stream. The base of the knoll consists of alternating thinly bedded sandstones and silty slates. The sediments are overlain by the second tuff of the Crafnant Volcanic Group which here begins with a massive, columnar-jointed, grey-weathering tuff. This passes up into a pumice-lapilli tuff in which the lapilli are flattened in a plane parallel to the base of the tuff. This is a mesoscopic eutaxitic structure and is usually a good indication of a welded tuff. However, Howells *et al* (1973) record crinoid fragments in this second tuff and do not recognise a welded fabric in any of the three tuffs of the Lower Crafnant Volcanic Group.

Follow the base of the tuff around the hinge of the syncline to Waenhir at [737584]. The tuffs appear eutaxitic and are columnar-jointed. The exposures are clearly on the southeastern limb of the syncline because the eutaxitic structure dips at 60° towards 312°.

Part 2: Waenhir to Llyn Bychan

Walk north obliquely across the strike of the second tuff unit towards a prominent crag with a broken down low wall leading to it. The second tuff is completed here by a columnar-jointed, structureless vitric tuff.

Continue northeast across a grass-covered depression about 50 m wide to a third tuff. Tuffaceous slates, siltstones and fine-grained sandstones are exposed at the base of the crag running along the northern side of the

Figure 57. View looking northeast from Clogwyn Mawr. I, II and III are the rhyolitic tuffs in the Lower Crafnant Volvanic Group. The approximate positions of the axial traces of an anticline and a syncline are shown.

grass-covered depression. The third tuff begins with 10 m of massive, crudely block-jointed, unbedded pumice-lapilli tuff containing occasional blocks or bombs of pumice. The rock is cleaved and lapilli are flattened in the plane of the cleavage. It is overlain by about 20 m of lapilli tuff which is columnar-jointed in the lower 10 m and is eutaxitic near the base. This is succeeded by a few metres of bedded crystal and lapilli tuffs which are sometimes nodular and may show soft-sediment deformation.

Howells *et al* (1973), on the basis of a detailed study of the three Lower Crafnant Volcanic Group and other tuffs, suggested that the tuffs are the result of sub-marine volcanism and were emplaced by water-suspended turbid flow. The evidence in support of the suggestion is very strong, despite the apparent eutaxitic fabric to be seen in some of the exposures.

Continue the traverse northeast. The third tuff is immediately overlain by 10 m of thinly bedded and laminated tuffaceous sands showing cross bedding. Horizons of carious weathering suggest the presence of carbonate-rich bands. There is then a small exposure gap before an old fence and a ruined wall, beyond which are crags of white-weathering, finely columnar-jointed, laminated rhyolitic tuff. The rock is mainly a flinty vitric tuff but some crystal- and lapilli-rich layers are present. It is strongly cleaved and folded on a mesoscopic scale. The tuffs now belong to the Middle Crafnant Volcanic Group.

Continue northeast across the stream and up the hillside to the summit at [742594]. A variety of assorted bedded tuffs and tuffaceous sediments are exposed on the flanks of the hill. Near the summit, a handsome cleaved rhyolitic vitric tuff contains blocks and bombs of pumice and vesicular rhyolite up to 0·5 m across. All sorts of bedded tuffs are present, but the dominant type is a dark blue, grey-weathering slaty tuff rich in crystals and lithic fragments. The tuffs and tuffaceous sediments were water-lain. A southeast–northwest traverse of the hill demonstrates that it lies astride the axial trace of the syncline, the hinge of which was noted earlier at Clogwyn Mawr.

Traverse due east across the bedded tuffs and tuffaceous sediments of the Middle and Upper Crafnant Volcanic Group for 1 km to Llyn Bychan. On the southeastern shore of Llyn Bychan the cutting at the side of the Forestry Commission track exposes cleaved, crystal-vitric tuffs and tuffaceous slates. The slaty rocks contain deeply weathered bombs of basic pumice now reduced by weathering to a soft, earthy condition. The clasts, flattened in the plane of the cleavage, are up to 30×50 cm.

Part 3: Llyn Bychan to Llyn Geirionydd

About 100 m northeast along the track from Llyn Bychan, the slaty beds with basic pumice bombs are overlain by blue crystal-vitric tuffs which show a poor development of columnar jointing. Continue around the bends in the track until the crystal-vitric tuffs are overlain by grey- and brown-weathering, cleaved, silty slates well exposed in the track side cutting. The bedding stripe is readily seen on the cleavage surfaces and it is clear that the axial trace of an anticline was crossed during the traverse to Llyn Bychan because the beds now dip southeast.

Leave the forest roadway by an indistinct footpath leading north to a stile over the boundary fence at [756596]. Walk northeast down the ridge of slates for 150 m to an old quarry in a wood. The slates are well exposed and show development of pyrite concretions along certain bedding planes. These are the Llanrhychwyn Slates which, when bedding and cleavage coincide, yield a graptolite fauna indicating the zone of *D. clingani*. Parallel lamination is the only sedimentary structure visible in these sooty black pyritous slates, which accumulated under quiet conditions beneath the reach of wave base.

Follow the old tramway from the quarry down to Llyn Geirionydd. Turn southeast and take the road to [766599]. Turn northeast along the track leading to the old mine and thence to Pen-y-Drum. The mine dumps are reached after 300 m. Very good specimens of blende and well formed pyrite, dolomite, haematite and quartz are readily turned up.

Continue to the top of Pen-y-Drum at [771609]. 400 m to the west a synclinal axial trace trends more or less north–south and, although the ground is largely grass-covered, the synclinal nature is quite obvious. Ashy slates dipping west at about 30° are sporadically exposed on Pen-y-Drum.

Walk northwest, down the dip, into a depression. Cross a rough, low wall running north–south along the depression onto ice-smoothed exposures of parallel-laminated, blue-black pyritous mudstones: the Trefriw Mudstones. The bedding is more or less horizontal and we are clearly in the hinge of the syncline. The Trefriw Mudstones are overlain in the core of the syncline by a small thickness of grey-black silty slates, the Grinllwm Slates. The slates are obviously bedded and contain rare lapilli of rhyolite up to 3 cm across which can be broken along the bedding, and yield poorly preserved diplograptids at this locality. They are probably Ashgillian in age.

Continue across the syncline to the top of the slope down to Llyn Geirionydd. The blue-black Trefriw Mudstones crop out here on the western limb of the syncline where bedding now dips at 32° towards 102°. The mudstones are cut by a close fracture cleavage dipping 52° towards 305°.

Walk northeast along the eastern edge of the plantation and continue to the road at Pen-yr-allt. Take the footpath up onto Grinllwm where the well bedded Grinllwm Slates occupy the core of an obvious syncline. The dark grey slates contain silty horizons and small, sporadic clasts of rhyolite up to 3 cm across.

The quiet, deeper-water conditions which prevailed during the accumulation of the Llanrhychwyn Slates continued during the deposition of the Trefriw Mudstones. The ashy nature of some of the sediments indicates not-too-distant volcanic activity. The lithology of the succeeding Grinllwm Slates also suggests accumulation mainly under quiet conditions, but the occasional siltstone and very fine-grained sandstone laminae may be the result of far-travelled turbidity currents. The occurrence of widely isolated, lapilli-sized clasts of rhyolite is more difficult to explain. They are too widely isolated to have been propelled directly, so perhaps some form of rafting may be invoked?

Having examined the slates preserved in the syncline, either follow the footpath northeast to Trefriw or return southwest to Pen-yr-allt, thence to the Monument at the outlet from Llyn Geirionydd, and then by way of the footpath to the road alongside the Afon Crafnant. Take the road and then the footpath back to Capel Curig.

EXCURSION 28. ABER TO LLYN ANAFON TO CWM YR AFON-GÔCH TO ABER

The excursion crosses the northeastern closure of the Arfon Anticline and lies mainly in a southeasterly dipping Lower Ordovician sequence in the southeastern limb of the maximal structure. Several large intrusions occur in the southeast of the ground covered by the

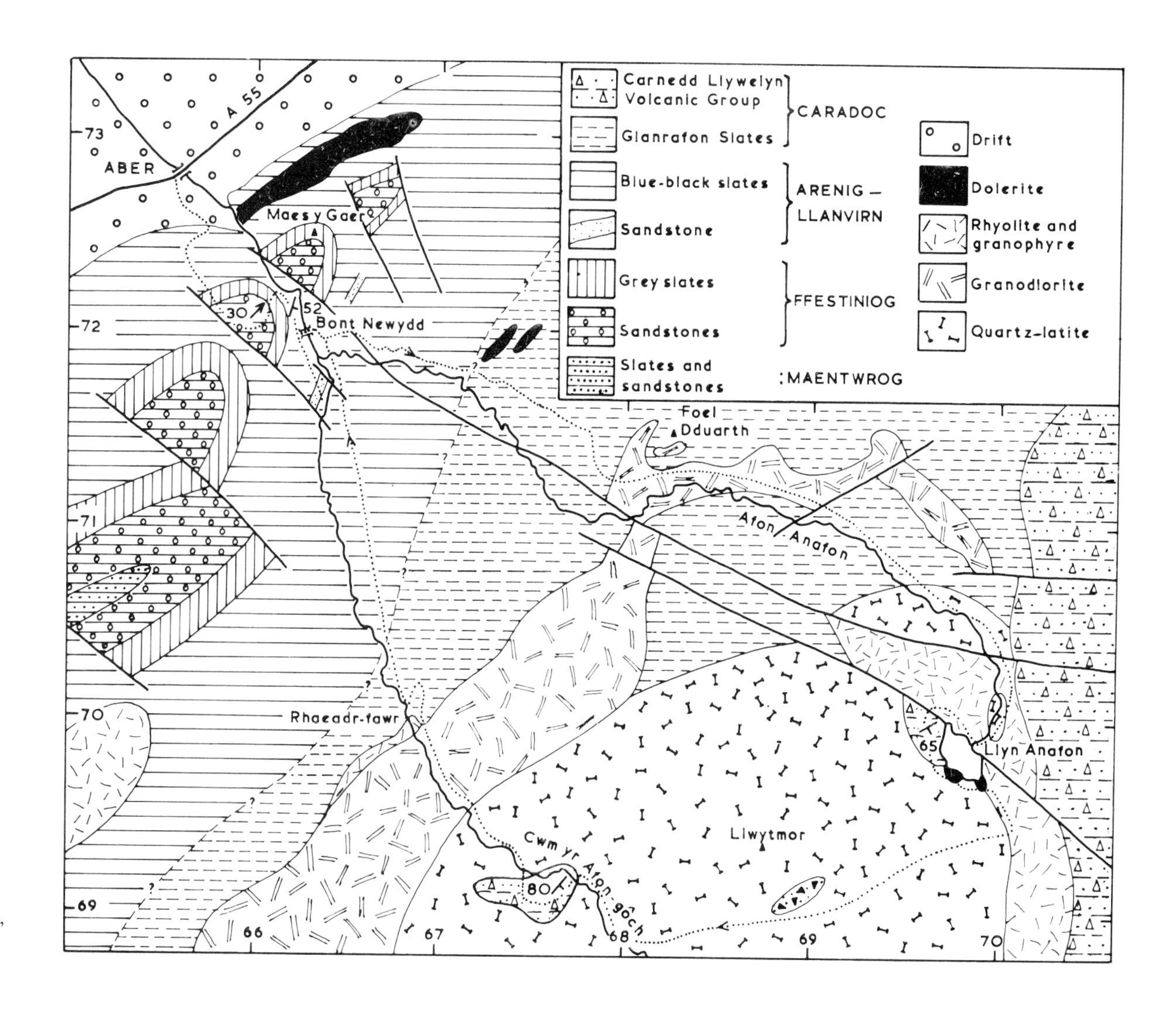

Figure 58. Excursion 28. Largely after Evans (1968), with permission, and Davies (1969).

traverse. The succession is given in figure 5 (column 7).
Use the excursion geological map, figure 58.

A car can be left at Aber or in the small car park at Bont Newydd [662720].

References: There is very little published on this area. The most useful pieces of work are two unpublished PhD theses by R A Davies (1969; Aberystwyth) and C D R Evans (1968; Aberystwyth).

1:50 000 OS sheet No. 115; or 1:25 000 sheets Nos SH66, 67, 76 and 77.

Part 1: Aber to Bont Newydd

Examine small and sporadic exposures of blue-black silty slates and mudstones on the hill slopes immediately south of Aber. Several footpaths lead onto the hillside. It is not possible to determine the attitude of the beds here, but climb up the hill to a vantage point at about [658720] above the wood overlooking the gorge of the Afon Aber. Look northeast towards Maes y Gaer and, on the opposite side of the gorge, see the fold in grey Ffestiniog sandstones. A tight, upright anticline is superbly exposed in vertical section in the wood beneath Maes y Gaer. The fold is much better seen in winter, when the trees have lost their foliage, than in summer.

The Ffestiniog sandstones are also exposed just below this viewpoint, and an anticlinal hinge plunges northeast down the slope at 30°, into the wood. Descend to the exposures just above and in the wood. The sandstones are massive and some are graded. Others have ripple-marked bedding planes. On Maes y Gaer large-scale cross bedding can be seen, and *Skolithos* and *Cruziana* can be found. Well sorted, coarse-grained sandstones are the dominant type, but granule conglomerates and medium- and fine-grained sandstones are also present. The sandstones probably accumulated in a sub-littoral environment. They are overlain by greyish blue silty slates, probably also of Ffestiniog age, but probably indicative of deeper-water conditions.

Continue down through the wood to the road in the bottom of the gorge. Turn south and, on the roadside 100 m north of Bont Newydd, examine an outcrop of finely laminated, black, pyritous siltstones. These beds overlie the grey-blue slates of presumed Ffestiniog age. Bedding dips 52° towards 108°, since the locality is in the southeastern limb of the Arfon Anticline, and about 100 m southeast of the axial trace. No fossils have yet been obtained from these siltstones but they are probably Arenig.

Continue south to Bont Newydd but do not cross the bridge. Instead, carry on along the footpath on the west bank of the river. Just beyond the confluence of the two streams, a feature above the footpath is formed by a sandstone about 15 m thick. The sandstone is coarse-grained at the base and contains discontinuous lenses and laminae of blue-black mudstone. It becomes finer-grained and more uniform upward and is burrowed. Near the base it contains chemogenic phosphatic nodules formerly termed *Bolopora undosa*, and is regarded by many as indicative of the basal Arenig sandstone. The sandstone is not the oldest member of the presumed Arenig succession however, for it occurs about 50 m above the base. The parallel-laminated, blue-black slates, silty slates and siltstones accumulated under tranquil conditions beyond the reach of wave base, but a period of shoaling is indicated by the bioturbated sandstone.

Part 2: Bont Newydd to Llyn Anafon

Return to Bont Newydd, cross the bridge and take the road eastsoutheast for 1500 m. At the end of the metalled road, take the footpath leading along the side of the valley of the Afon Anafon. At Foel Dduarth, the spur from Foel Ganol, strongly cleaved, pink- and grey-weathering felsite is exposed. It is emplaced in mildly contact-altered, splintery green slate with chloritic spots. The slates are regarded as Caradoc in age by Davies (1969), although no fossils have been obtained. The contact is exposed on the hillside and can be seen to be irregular, as at [681715], and near-horizontal.

Continue along the track until the river below flows over bedrock at [697706]. The rock is a quartz–latite and consists of feldspar phenocrysts in a blue-grey microcrystalline base. The outcrop is probably a faulted marginal facies of the very extensive mass to the southwest on Llwytmor.

About 150 m east of Llyn Anafon, flow-banded and flowage-folded rhyolite, presumably intrusive, crops out. The intrusion extends south from Llyn Anafon for 1200 m up towards the ridge of Foel Fras. Cross to the northern end of Llyn Anafon and examine the exposures on the western slopes leading down to the lake. The rocks exposed are andesitic lavas and tuffs, dacite tuffs, agglomeratic tuffs and green slates. Bedding dips at 65° towards 150°, but the relationships to the surrounding quartz–latite mass have not been worked out. It seems that the quartz–latite is a sill-like mass and that these tuffs, included by Evans (1968) with the Carnedd Llywelyn Volcanic Group, are exposed in an erosional window through the sill-like mass. Continue southeast around the western shore of the lake and, just before the point where a stream flows northeast into the lake, the volcanics are seen to be overlain by a sill-like dolerite intrusion. The dolerite crops out again at the southern end of the lake, where a larger stream flows over a small waterfall and enters the lake.

The junction of the quartz–latite against the intrusive rhyolite can be seen at [700696] to dip steeply eastward. Blocks of rhyolite can be found in the marginal facies of the quartz–latite, indicating that the quartz–latite is younger than the rhyolite.

Walk southwestward towards a col on Llwytmor, across excellent exposures of quartz–latite. In the hand specimen the rock consists of a grey-green microcrystalline base supporting feldspar phenocrysts and black ferro-magnesian microphenocrysts. The ferro-magnesian minerals include both hornblende and pyroxene. Descend westward into Cwm yr Afon-gôch.

Walk downstream to a point [676692] 50 m upstream from the first rowan tree. Cleaved, unbedded, rhyolitic tuffs are exposed on the south bank of the stream whereas columnar-jointed quartz–latite crops out on the north bank. Further exposures of volcanic rocks can be examined south of the stream, where they are seen to include bedded andesitic and rhyolitic crystal tuffs and pumice tuffs. Bedding dips at 80° towards 130°. Evans equated the volcanic rocks with those exposed at Llyn Anafon, grouped them with the Carnedd Llywelyn Volcanic Group, and regarded them as exposed in an erosional window and as constituting evidence for the sill-like nature of the quartz–latite intrusion.

Continue downstream across reasonably well exposed, columnar-jointed quartz–latite onto poorly exposed microgranodiorite at about [672695]. From here, look southwest and south at the hillside. The approximate junction of the columnar-jointed and reasonably well exposed quartz–latite against largely grass-covered microgranodiorite can be seen to be inclined at about 60–70° to the northwest.

Continue downstream across sporadic exposures until, about 100 m southeast of Rhaeadr-fawr (the Aber Falls), extensive pavements of medium-grained granodiorite with abundant xenoliths can be examined. The xenoliths are of two types: sedimentary xenoliths which are small, ranging up to 1 cm, and cognate xenoliths of an early, basic facies ranging up to 10 cm across.

Descend the cliff by way of the footpath on the east and examine the exposures at the foot of the falls. Hornfelsed laminated pelites of presumed Lower Ordovician age are exposed against a marginal facies of the microgranodiorite. A lens or small raft of sediment has been incorporated within the marginal facies, and another can be seen about 10 m up the scree to the east. The contact, although complicated in detail, is more or less vertical and the microgranodiorite is presumably a thick, dyke-like mass.

Return by way of the footpath across largely drift-covered ground to Bont Newydd and Aber.

EXCURSION 29. PENMAEN-MAWR TO TAL-Y-FAN TO LLANFAIRFECHAN

The aim of this excursion is to examine the Penmaen-mawr intrusion and the Caradocian sequence of sediments and volcanics into which it is emplaced. The succession is given in table 5.

A car can be left for the day in Penmaen-mawr.
Use the excursion geological map, figure 59.
References: Davies (1969) and Stevenson (1971).
1:50 000 OS sheet No. 115; or 1:25 000 sheets Nos SH67 and 77.

Part 1: The Penmaen-mawr Intrusion

Take the footpath onto the hillside from Graig-lwyd [719785] and head south obliquely up the hill towards the obvious contact of the intrusion with the country rock at [718752]. The country rock consists of blue-black and grey-black slates and silty slates and, near the contact, the cleavage strikes north–south and dips at 70° east. Away from the intrusion the cleavage swings to more or less east–west. Notice the

Table 5. Terminology used in describing the Caradoc sequence southeast of Penmaen-mawr†.

Davies (1969)	This excursion	Stevenson (1971)
Bedded Pyroclastic Series	Crafnant Volcanic Group — Basic tuffs	Crafnant Volcanic Group — Basic pumice tuff
Upper Glanrafon Beds: Mudstone with intercalated rhyolitic tuffs Sandstones	Upper Glanrafon Beds: Slates with intercalated rhyolitic tuff Sandstones	Glanrafon Beds: Slate and thin rhyolitic tuff Sandstones
Capel Curig Volcanic Suite	Capel Curig Volcanic Group: Rhyolitic tuffs	Conway Volcanic Group: Rhyolitic tuff in Maen Amor Division
Lower Glanrafon Beds: Sandstones Drosgl Volcanic Suite Slates Manganiferous tuffs Slates	Lower Glanrafon Beds: Sandstones Carnedd Llywelyn Volcanic Group Blue-back silty slates Manganiferous tuffs Blue-black silty slates	Conway Volcanic Group (cont.): Sandstone in Maen Amor Division Andesitic tuff in Gyrach Division ‡ Slates below the Conway Volcanic Group

† The successions lie entirely within the Caradoc.
‡ The base of the Conway Volcanic Group remains obscure because Stevenson has interpreted his 'Lower Banded Rhyolites' as lavas, whereas Davies and the author prefer to regard the formation as intrusive rhyolite.

absence of any obvious contact metamorphic phenomena. The marginal facies of the intrusion is a very fine grained microtonalite. It is a light grey, microcrystalline rock with a sub-conchoidal fracture. It carries small phenocrysts of plagioclase and black ferro-magnesian microphenocrysts which, in thin section, are seen to be augite and enstatite. The contact is steep to vertical.

Enter the old Graig-lwyd quarry and notice that the grain size of the matrix coarsens inward from the contact. Veins of a more acid facies cut the tonalite and irregular segregations of the same material occur within the tonalite. The intrusion appears to be a vertical, plug-like mass and the way in which the cleavage in the slaty country rock swings around the intrusion suggests it pre-dates the folding of the rocks. It is probably mid-Ordovician in age.

Part 2: Graig-lwyd to Tal-y-fan

Walk eastward for 500 m to [723751] where a stream has cut a section about 100 m long in the Caradocian country rock. The lowest exposed rocks are blue-black siltstones and mudstones dipping at 40° towards 100°. Cleavage is vertical and strikes east–west. They are succeeded by about 30 m of purple lapilli tuffs of probable andesitic composition, which are overlain by about 30 m of finely laminated, green mudstones, and then by about 70 m of bedded andesitic lithic tuffs with thin green mudstone partings. The tuffs are rich in manganese and have been called 'manganiferous tuffs' by Davies. The highest beds exposed in the section are blue-black siltstones and mudstones. The mudstones and silt-stones intercalated with the bedded andesitic tuffs show only parallel

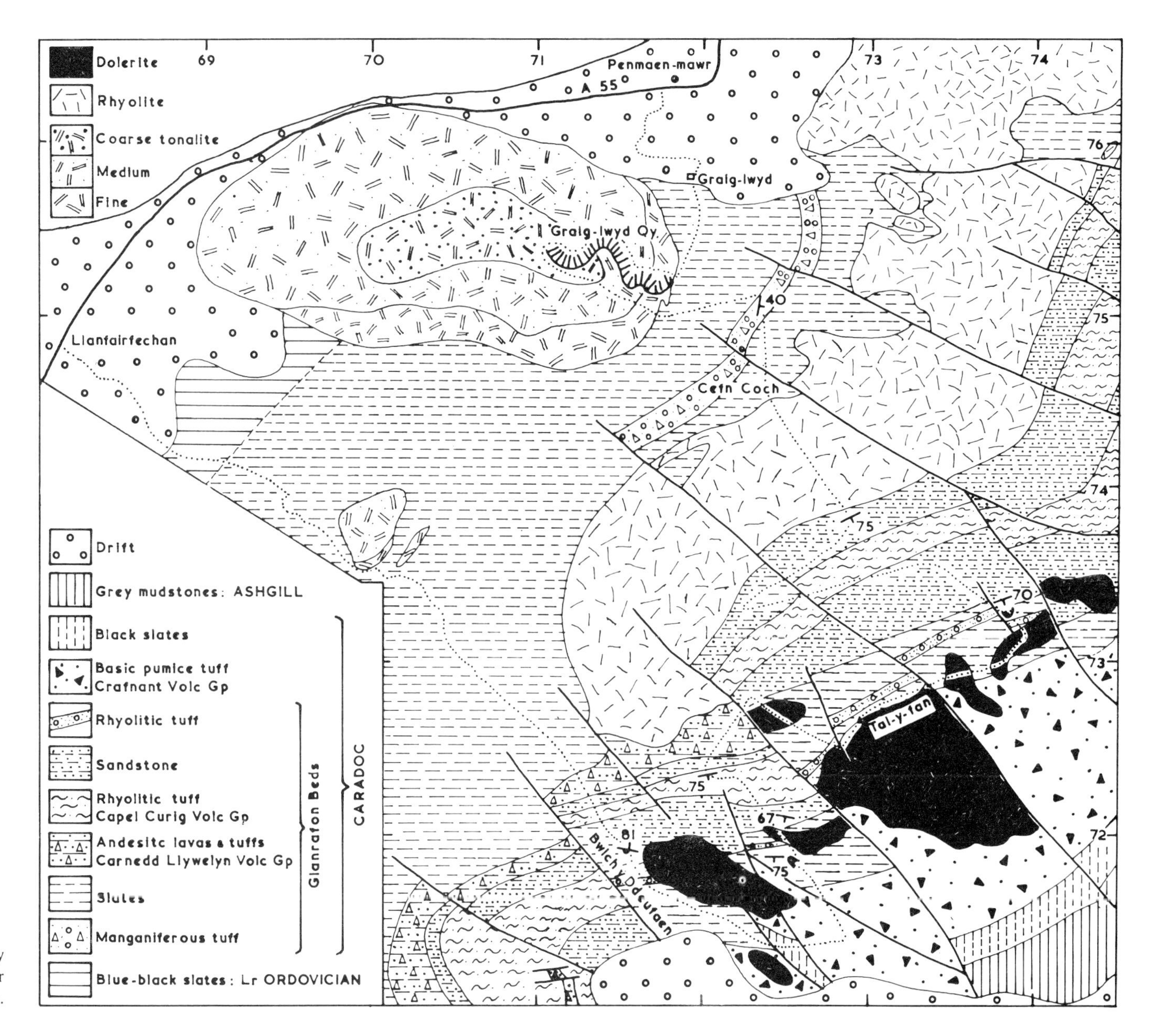

Figure 59. Excursion 29. Largely after Davies (1969) and partly after Stevenson (1971), with permission.

lamination. The sequence probably accumulated under quiet marine conditions beneath the reach of wave base.

Walk south up onto Cefn Coch where, at [724746], flow-banded rhyolite is exposed. The rock is uniformly banded, the bands dipping at 70° towards 120°. Stevenson (1971) regarded this rock as part of a group of rhyolite lavas called the 'Lower Banded Rhyolites' which constitute the oldest member of his Conway Volcanic Group. Davies (1969), however, regarded the rhyolite as an intrusive sill-like sheet, and it is this interpretation which is favoured by the author.

Continue southeast for about 1 km to [729738], where coarse-grained, greenish sandstones are exposed. The sandstones are rich in tuffaceous material, including feldspars and rhyolite fragments. They are massive, although they show either parallel lamination or cross bedding. The strike has swung clockwise so that bedding now dips at 75° towards 155°. The sandstones are regarded as Lower Glanrafon Beds by Davies (1969) and as members of the Conway Volcanic Group by Stevenson (1971). Cross southwestward to [722733] where the sandstones, which are conglomeratic and coloured purple, show trough cross bedding. The sediments are essentially volcanic sands and their sedimentary characteristics suggest accumulation in shallow water under the influence of strong currents. Shoaling has therefore occurred and the environment was probably sub-littoral.

Contour eastward for 1500 m across poorly exposed ground to [734736] where a rhyolitic pumice-lapilli tuff overlies the sandstones. The tuff is welded and is continuous with the Capel Curig Volcanic Group away to the southwest. Stevenson (1971) regarded this tuff as lying within his Maen Amor Division of the Conway Volcanic Group. The welded character of the tuff indicates the shoaling culminated in emergence and the tuff was emplaced subaerially.

Continue southeast to the old slate quarry at [738733]. On the way, note the sporadic exposures of southeasterly dipping sandstone which both Davies and Stevenson regarded as members of the Glanrafon Beds. The sandstones are green and tuffaceous, and contain pebbly horizons with rhyolite and quartz clasts. Badly preserved brachiopods and trilobites of Soudleyan age can be obtained from the sandstones. The fossiliferous sands were clearly laid down under shallow marine conditions and so the period of emergence, during which the welded tuff accumulated, was probably of short duration.

At the slate quarry the rocks exposed are silty, blue-grey slates with occasional Soudleyan brachiopods.

Head southwest up the spur on to the ridge of Tal-y-fan.

Part 3: Tal-y-fan to the Unnamed Hill at [720723]

The ridge of Tal-y-fan is formed by a thick dolerite sill dipping southeast. Towards the middle of the sill the texture becomes gabbroic. In thin section the rock is seen to contain both prehnite and pumpellyite, mainly after plagioclase. Walk southwest along the ridge to the summit and then northwest to exposures of a rhyolitic crystal-vitric tuff. The rock is crudely bedded and greenish grey in colour. It is described by Stevenson as having a eutaxitic texture in thin section, although it has none of the obvious characters of a welded tuff in the field. It lies within the Glanrafon Slates.

Continue northwest across the stream and a fault to about [722728] and turn to walk up the hillside to the summit at [720723]. The lowest exposures are of a rhyolitic, crystal-vitric, lapilli tuff overlain by a flow-banded, flowage-folded and strongly amygdaloidal andesite. The amygdales are of chlorite. This is succeeded by a flow-banded and flowage-folded rhyolite which is possibly intrusive. They are followed by a cleaved, andesitic, crystal-lithic lapilli tuff. The lapilli are of rhyolite. The rock is green but weathers to a khaki shade. It is overlain by a red-purple-weathering, andesitic agglomeratic tuff in which the clasts of rhyolite and andesite range up to 5 cm across. Davies (1969) regarded the volcanic rocks up to this horizon as comprising the northernmost members of the Drosgl Volcanic suite, a suite that is equivalent to the Carnedd Llywelyn Volcanic Group.

The traverse then crosses an exposure gap which is believed to conceal Glanrafon sandstones. The succeeding exposures are of a cream-weathering, porcellanous rhyolite or rhyolitic tuff. The rock is strongly cleaved and is succeeded by a white-weathering, eutaxitic, rhyolitic, pumice-lapilli tuff. The eutaxitic structure dips at 75° towards 170° and is parallel to the upper and lower surfaces of the sheet of tuff. The eutaxitic tuff, which shows a nodular development just north of the summit, is a member of the Capel Curig Volcanic Group.

Part 4: Summit [720723] to Bwlch y Ddeufaen

Follow the wall east from the summit across the rhyolitic tuffs, and then southeast to an area which has been fenced off to keep out animals. Excellent lichen-free exposures of coarse-grained sandstones can be examined. Pebble beds, mud-pellet-rich horizons, cross bedding and washouts can be studied. Bedding dips at 67° towards 188°, whereas cleavage dips 80° towards 007°. These are the largely volcanic Glanrafon sandstones overlying the Capel Curig Volcanic Group. This is a good point from which to look back at the dolerite sill of Tal-y-fan and note its slightly transgressive nature, since the sill is seen to dip less steeply south than the bedding.

Continue southeast, across green and purple mottled, silty slates of the Glanrafon Beds, to a coarse-grained dolerite sill. Continue downhill across siltstones and fine-grained sandstones and then, just above an old slate quarry at [725718], across a prominent pink-weathering, coarse-grained feldspathic sandstone dipping at 75° due south.

Descend to the slate quarry where blackish grey, silty slates contain abundant fragmentary trilobites and small brachiopods of Soudleyan age. The dumps are particularly fruitful. The slates are overlain by a thin, coarse-grained sandstone which brings the Glanrafon succession to a close.

The Glanrafon Beds are overlain downhill by a strongly cleaved, greenish-blue, bedded, basic crystal-lithic tuff. There seems to be no reason to consider the tuff as other than sub-marine in its accumulation. The tuff is regarded by Davies (1969) as equivalent to the Bedded Pyroclastic Group of Central Snowdonia but as the lowest local member of the Crafnant Volcanic Group by Stevenson (1971): this grouping of Stevenson's seems preferable to the author. The tuff is of additional interest because some thin sections show shards replaced by prehnite, albite and rare pumpellyite. The rocks are therefore in the prehnite–pumpellyite facies of very low grade regional metamorphism.

Continue down to the road. Turn west and follow it to Bwlch y Ddeufaen. Follow the higher of the two tracks to [715719] where, at the base of the last pylon before the bend in the line of cables, an excavation exposes highly fossiliferous, fine-grained Glanrafon sandstones. The abundant brachiopods are sometimes deformed and can be used to estimate the strain the rocks have experienced. The fauna indicates a Soudleyan age. The beds do not appear to have been affected by hill creep, yet are clearly overturned, dipping at 87° towards 010°. This attitude is probably the result of an F_3 cross fold, the hinge of which is clearly delineated on the excursion map immediately to the southwest of Bwlch y Ddeufaen.

Follow the path northwest down the valley of the Afon Ddu to Llanfairfechan, and so back to Penmaen-mawr.

EXCURSION 30. THE AREA AROUND CONWY

The aim of this excursion is to examine the Conway Volcanic Group in its type locality and to examine the overlying sediments of the Ashgill, Llandovery and Wenlock Series. The succession is given in figure 5 (column 9).

Use the excursion geological map, figure 60.
References: Elles (1909) and Stevenson (1971).
1:50 000 OS sheet No. 115; or 1:25 000 sheet No. SH77.

Part 1: Penmaen-bach

It is undesirable, and in any case just about impossible, to park on the main A55 road. It is usually possible to park if you turn north over the railway bridge at [761784]. This leaves you with a walk along the road of about a kilometre.

As you approach the Penmaen-bach igneous mass, the excellent columnar jointing becomes evident. The attitude of the columns suggests

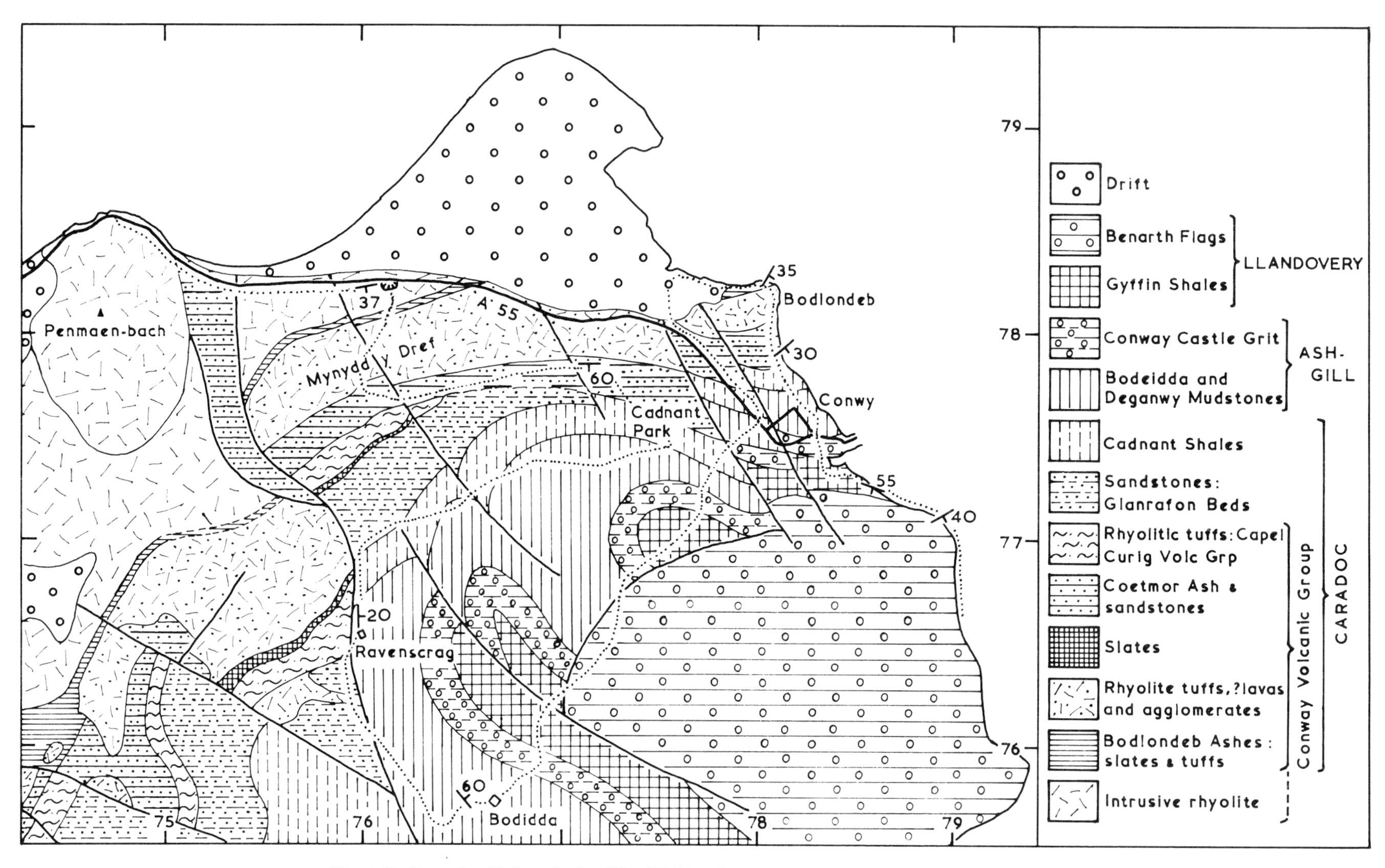

Figure 60. Excursion 30. Largely after Elles (1909) and partly after Stevenson (1971), with permission.

the mass is a vertical, cylindrical intrusion. The rock can be examined at the roadside and on the foreshore. It is a flow-banded and flowage-folded rhyolite which, in the hand specimen, is feldspar-porphyritic and the matrix consists of microcrystalline quartz and feldspars.

Part 2: Mynydd-y-Dref (Conwy Mountain) to Bodidda

Return to the quarry on the south side of the road near the bridge over the railway [761783].

The rock exposed is a flow-banded rhyolite, greenish blue on fresh surfaces, but weathering to brown. Individual flow bands are continuous over several metres. The bands swirl in very broad arcs, but obvious flowage folds are not developed. Some bands are brecciated. The banding dips at 37° towards 170°, which suggests the mass is concordant. These rocks were regarded as rhyolite lavas by Elles and Stevenson, but Davies (1969) regarded their lateral equivalents to the southwest as intrusive. An intrusive origin is also preferred by the author. The mass is thought to have a sill-like form.

Climb up the hill around the western side of the quarry and across a slight break in slope which probably conceals the Bodlondeb Ashes. As the slope steepens again the rocks exposed are brecciated rhyolites, which may or may not be intrusive. They belong to the Upper Brecciated Rhyolites of Elles; Stevenson groups their lateral equivalents as tuffs in the Gyrach Division. Further up the hillside the rocks consist of blocks and bomb-like masses of rhyolite up to 0·5 m across, in a clastic rhyolitic matrix. At a higher level blocks of rhyolite up to 1 m across are held in a matrix of flow-banded and flowage-folded rhyolite. The rocks are difficult to interpret. They may represent agglomerates and lavas, or they may be entirely intrusive. The author suspects they are agglomerates and lavas. The summit of the hill consists of flow-banded and gently flowage-folded rhyolite. Brecciation is absent. These rocks are probably intrusive. A strongly nodular horizon is developed a little way down the dip slope from the summit and is well seen between the two east–west footpaths. The rocks are very rich in ferric oxides and are cavernous, indicating that leaching of formerly abundant sulphides has taken place.

Continue down the path towards Conwy and at [769778] examine the obviously bedded rocks exposed in the old quarries. The beds dip 60° due south and are cut by a small reversed fault with a throw of 2–3 m. In the hand specimen the rocks can be seen to be coarse-grained, grey, quartz–feldspar sandstones. They make up the Coetmor Ash Group of Elles and are placed in the Maen Amor Division by Stevenson. They complete the Conway Volcanic Group succession.

The overlying rocks are the Cadnant Shales which are sporadically exposed in Cadnant Park. A footpath leads from the quarries to the road and then into Cadnant Park. Black shales are poorly exposed near the footbridge and have yielded a fauna indicating the zone of *D. clingani*. (The best section is in the railway cutting immediately northwest of Conwy Station but this, of course, is normally not accessible.) The sooty black shales show a fine parallel lamination. Lithologically similar shales and slates with an identical graptolite fauna can be traced across Gwynedd from Conwy in the east to Llanbedrog in the west, and it is evident that the sediments accumulated under quiet conditions beneath the reach of wave base.

Ashgillian mudstones, the Bodeidda and Deganwy Mudstones, overlie the Cadnant Shales, but they are very poorly exposed in the Cadnant area. Take the Sychnant road west and southwest to Ravenscrag where, at [760766], small brachiopods and trilobite fragments can be obtained from grey mudstones of Ashgillian age. Notice that the strike has swung and that bedding now dips at 20° towards 087°. Continue south towards Bodidda [767757], where the mudstones are again seen in small roadside exposures. Bedding here dips at about 60° towards 050° and we have therefore crossed a synclinal structure occupying the ground between Cadnant Park and Bodidda, and the fold plunges towards about 120°. This southeasterly plunging syncline, which is very clearly delineated on the excursion map, constitutes a large-scale F_3 structure.

Return along the road to Conwy and carry on to Bodlondeb.

Part 3: Bodlondeb to Conwy Castle

Take the footpath which runs first east and then south along the shore around Bodlondeb Wood to [780782].

The lowest beds exposed at the side of the path belong to the Bodlondeb Ash Group. The succession begins with a massively bedded, agglomeratic and lapilli-rich, crystal-lithic tuff of andesitic composition. Bedding is difficult to measure, but appears to dip at 35° towards 120°. The rock is feebly cleaved, and cleavage dips at 60° towards 020°.

The pyroclastic horizon is overlain by a purple and blue porphyritic andesite lava and, in turn, by a magnificent andesitic lapilli tuff containing occasional bombs of amygdaloidal andesite. The amygdales are of chlorite and calcite. This is overlain by 1–2 m of blue-black, laminated shale from which Elles obtained graptolites including *G. teretiusculus*. The shale is succeeded by a crystal-rich, andesitic tuff about 0·5 m in thickness.

The succeeding beds are orange-weathering and rhyolitic in composition, and appear to be massively bedded, clastic rocks—a tuff of some kind. The rocks have a close but irregular parting sub-parallel with the massive bedding, but pumice lapilli are absent. The beds have been placed by Elles (1909) in her Upper Brecciated Lava Group. The rocks become columnar-jointed and cream-weathering towards the top and breccia horizons are present. The sheet may be an ignimbrite but, whatever its origin, appears to dip at 37° towards 120°, in accordance with the enclosing sediments.

The overlying beds are quartz–feldspar sandstones, coarse-grained and much stained by ferric oxides. The beds have been mapped as the Coetmor Ash Group by Elles and exposures persist, especially along the foreshore, to within 100 m of the town wall.

The Cadnant Shales and Bodeidda Mudstones are sometimes exposed on the foreshore between here and the Castle, but at the present time they are largely concealed by gravel, sand and mud.

Conwy Castle is built on an outcrop of the Conway Castle Grits and, provided care is taken, they can be examined in the cutting on the Trefriw road. They consist of a succession of interbedded, calcareous greywackes and shales. Calcite fills joints and tension gashes. A prominent fault can be examined on the eastern side immediately before the arch at the castle wall. The greywackes are graded and range from 3 cm to several metres thick. Large mud pellets, always rounded and up to 10 cm across are common at the base of units. Sole markings include flute casts and linguloid ripples. The rocks have yielded a Hirnantian fauna and are therefore topmost Ashgillian. They are clearly turbidites.

Part 4: Conwy Castle to Benarth

Continue south for 100 m along the Trefriw road and then turn east along the road to the foreshore.

Exposures of the overlying Silurian Gyffin Shales can be examined just past the mussel-washing plant and also on the foreshore. On the foreshore the rocks are inter-laminated grey and black slates with a very good slaty cleavage; they are, in fact, slates rather than shales. The black laminae contain a graptolite fauna of Llandovery age. The parallel-laminated Gyffin Shales accumulated under quiet conditions beneath the reach of wave base; and the influx of turbidity currents, which earlier gave rise to the Conway Castle Grits, had now ceased. Over much of the exposed pavement of rock bedding dips at 55° towards 198°, whereas the slaty cleavage dips at 75° towards 165°, but occasional small-scale flexures are present and to these mesoscopic structures the slaty cleavage (S_1) bears an axial-planar relationship. These folds are therefore F_1 structures and plunge at 40–45° towards 242°. (It will be remembered that an F_3 syncline plunging towards 120° was crossed between Mynydd-y-Dref and Bodidda and the structure forms the most obvious feature on the accompanying excursion map, figure 60.)

Continue southeast along the foreshore. The Gyffin Shales crop out until about 200 m before the headland. At this point thin, fine-grained greywacke sandstones and greywacke siltstones appear within the slates. Calcareous concretions are common, and the thin black slaty laminae yield graptolites indicating the base of the Wenlock. The proportion of sandstones to slates increases towards the headland.

At the headland good greywacke sandstones enter in force. The sandstones are commonly graded and are often rich in mud pellets. Soft-sediment deformation, probably slump structures, are common. Sole markings include flute casts, load casts, groove casts and prod marks. The slaty interbeds yield Wenlock monograptids. The greywacke sandstones indicate a renewed influx of turbidity currents into the basin.

South of the headland the rocks have been folded, and folds plunge at angles ranging from 25° to 45° towards directions ranging from 090° to 120°. Bedding is sometimes seen to dip more steeply southward than the cleavage, and such beds are probably inverted, but since the beds have almost certainly been subjected to two strong phases of folding, bedding/cleavage relationships should be interpreted with caution.

In this excursion, the sequence from the Capel Curig Volcanic Group to the Ashgillian is examined. Attention is given to the problem posed by the presence of strongly developed northwest–southeast-trending structures in the ground between Dolgarrog and Conwy.

It is possible to find somewhere to leave a car for the day at Tal-y-bont. Use the excursion geological map, figure 61.
References: Davies (1969) and Stevenson (1971).
1:50 000 OS sheet No. 115; or 1:25 000 sheet No. SH76.

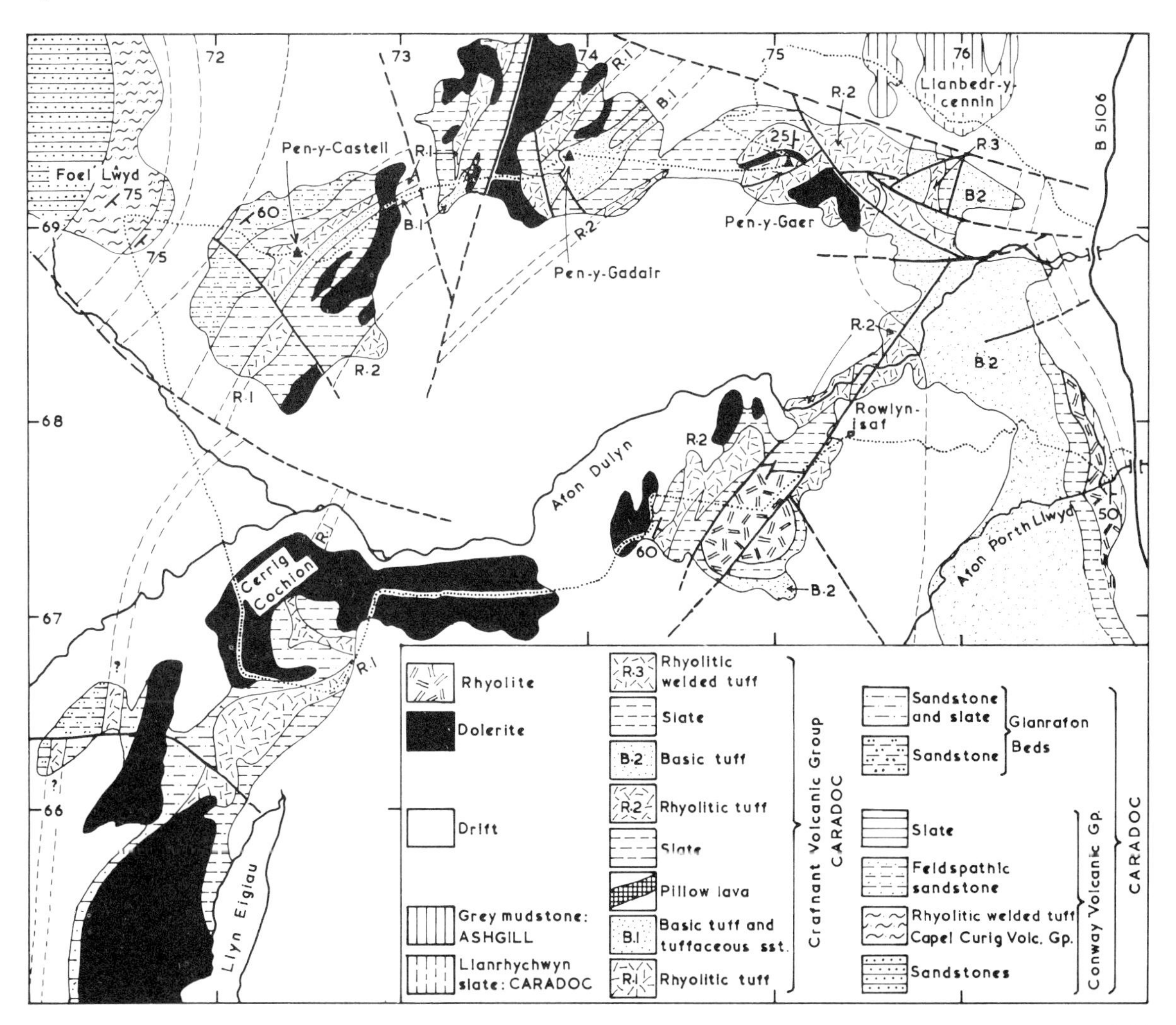

Figure 61. Excursion 31. Largely after Stevenson (1971), with permission.

Part 1: Porth Llwyd

Take the minor road from [769678] westward and then south to the Afon Porth Llwyd.

Here black slates, which may represent the Llanrhychwyn Slates, are exposed. No fossils have been obtained. Sooty laminae indicate that the beds dip at 50° towards 090° and they overlie the Crafnant Volcanic Group.

Cross northeast to the prominent crag at [768678]. The exposed rocks underlie the presumed Llanrhychwyn Slates, and were taken by Stevenson (1971) to represent the highest local member of the Crafnant Volcanic Group. The rock is a cream-weathering, flow-banded rhyolite, perhaps intrusive. It was regarded as a welded tuff by Stevenson (1971). The rhyolite, which is about 30 m thick, is underlain by a black tuffaceous slate. No fossils have been obtained from the slate. Continue northwestward up the hillside across exposures of agglomeratic basic tuffs which underlie the thin, black slate horizon. The tuffs are green and grey, crudely bedded rocks, rich in lapilli of basic pumice and lithic fragments as well as crystals. Occasional bombs of vesicular andesite are present.

The basic tuffs, slate and rhyolitic rocks comprise the Upper Division of the Crafnant Volcanic Group of Stevenson.

Part 2: Porth Llwyd to Cerrig Cochion

Take the track past Rowlyn-isaf and on to the hillside exposures on Waen Bryn-gwenith [750676].

Examine the fine crags leading to the summit of the hill. The rocks are rhyolitic and include clasts of black mudstone. They were regarded as tuffs by Stevenson (1971), but the rocks are amygdaloidal and flow-banded rhyolites. Field relations suggest they are intrusive. Walk due west for 500 m around the northern slope of the hill, across further rhyolite as well as tuff and interbedded tuffaceous slate, to a prominent outcrop immediately west of the track at [743676]. A coarse-grained dolerite with plagioclase-rich segregations is exposed here. The crude columnar jointing suggests a sheet dipping southeastwards and, at the track, the overlying slates and tuffs can be seen dipping at 60° towards 115°; the sheet is thus essentially concordant.

Take the track southwest for 500 m and then the footpath west towards crags of dolerite. Dolerite crags project from the drift at intervals before the crags of Cerrig Cochion are reached: these probably represent exposures of a single sheet. The rock is often sufficiently coarse-grained to be a gabbro. At the foot of Cerrig Cochion turn southwest to exposures beyond the head of the stream at [726666]. Rhyolitic pumice-lapilli tuffs and crystal tuffs are overlain by tuffaceous slates and underlain by black mudstones. The tuffs and overlying sediments are regarded by Stevenson as members of his Lower Division of the Crafnant Volcanic Group. Davies regarded the tuffs as the Lower Rhyolitic Tuff. The blackish mudstones beneath the tuff are grouped with the Glanrafon Beds by both authors.

Part 3: Cerrig Cochion to Foel Lŵyd

Walk northwest across the dolerite of Cerrig Cochion and then across 2 km of drift-covered ground to the prominent exposures at Foel Lŵyd above a complex of sheepfolds at [691715].

Part 4: Foel Lŵyd to Pen-y-Castell

Examine the exposures of rhyolitic tuff above the sheepfolds. The rocks are pumice-lapilli tuffs with a pronounced eutaxitic structure, dipping at 75° towards 137°, and which is parallel to the top and bottom of the ignimbrite sheet. They were clearly emplaced subaerially. They have been mapped by Davies as the Capel Curig Volcanic Group, and as tuffs within the Gyrach Division of the Conway Volcanic Group by Stevenson.

Descend to the exposures below the sheepfolds. The rocks are massively bedded, grey, clean, quartzo-feldspathic, coarse-grained sandstones. The rocks are strongly cleaved. The sands are shallow-water marine sediments and so, consequently, the subaerial episode during which the ignimbrite was emplaced was brief and was followed by submergence. The sandstones are placed by Stevenson in his Maen Amor Division of the Conway Volcanic Group; Davies maps them as Upper Glanrafon Beds.

Walk east across the peat-covered gap to the exposures on Pen-y-Castell. The lowest exposures are interbedded coarse- and fine-grained, greenish sandstones and siltstones. They are trough cross bedded and cross laminated. Soft-sediment deformation is common in the form of slumped foreset beds and convolute lamination. Occasional conglomeratic horizons consist of rhyolite pebbles in a tuffaceous matrix, and thin horizons of white-weathering, porcellanous vitric tuffs are also present. These sands are essentially volcanic in origin and many of the clastic lithic fragments are andesitic and rhyolitic. They are shallow-water deposits which were laid down under the influence of strong currents, probably in a sub-littoral environment. The sandstones are chopped into cigar-sized pieces by the intersection of two cleavages. The earliest, S_1, is a slaty cleavage in the finer-grained tuffaceous lithologies and dips 67° towards 318°, whereas the second cleavage *in this exposure* (but probably S_3 of the structural section, p 38) dips at 85° towards 056°. A suite of mesoscopic folds associated with this second cleavage is present plunging towards the southeastern quadrant. Davies and Stevenson each map these sandstones as members of the Glanrafon Beds.

The presence of a slaty cleavage (S_1), with Caledonoid trend deformed by a later set of mesoscopic structures (F_3), and an associated strong fracture cleavage (S_3) suggests to the author that the open folds plunging southeast, such as the syncline plunging southeast from Bwlch y Ddeufaen to Dolgarrog (figure 59), and the syncline plunging southeast from Mynydd-y-Dref near Conwy (figure 60), are also of F_3 generation.

About 100 m to the eastsoutheast, beyond the brow of the hill, there are jagged outcrops of grey, laminated siltstones and fine-grained sandstones. These outcrops also show two cleavages. The first, S_1, is a slaty cleavage in the finer-grained lithologies and dips 80° towards 322°. The second is probably S_3 and it is a very strong fracture cleavage dipping 87° towards 047°.

Carry on eastwards to exposures of overlying white-weathering, rhyolitic tuff. The rock is flinty and nodular. At Pen-y-Castell summit, the base of the tuff is seen to be unwelded, crystal-rich and strongly cleaved. This basal zone is overlain by a nodular facies above which the tuff is eutaxitic and welded. This is succeeded by a bedded pumice-lapilli tuff and the succession of tuffs ends with a rhyolitic vitric tuff. It seems likely that the basal tuff, the nodular horizon and the eutaxitic horizon represent successive ash flows and ash falls. In part, at least, they represent subaerial accumulations. Stevenson maps them as the Lower Division of the Crafnant Volcanic Group and equates them with the rhyolitic tuffs of Cerrig Cochion, whereas Davies regarded them as a local development within his Upper Glanrafon Beds and older than the Cerrig Cochion tuffs.

There is a small exposure gap above the tuff before coarse-grained overlying sandstones crop out. About 50 m across the strike, a prominent white-weathering, bedded, rhyolitic vitric tuff is interbedded with the sediments. Pebbles of rhyolite occur within the tuff. The beds are very similar to the volcanic sands seen earlier on this traverse.

Part 5: Pen-y-Gadair to Pen-y-Gaer

Traverse northeast along the strike of the sandstones to the bwlch where a prominent line of exposures can be seen. Examine the large *roche moutonée* of coarse-grained dolerite on the way. A fence runs northeast towards Pen-y-Gadair. The first exposures on the northwestern side of the fence are of rubbly and blocky vesicular basalt, whereas on the southeastern side of the fence the rock is clearly a basaltic pillow lava. Stevenson regarded the lavas as part of the Lower Division of the Crafnant Volcanic Group; Davies does not record them. Some 40 m further along the northwestern side of the fence a basic, agglomeratic, pumice-lapilli tuff crops out. Continue along the fence to the wall. On the other side of the wall are exposures of a superb pillow lava. Pillows range up to 2 m across and the structure of pillows, showing concentric shells alternately rich and deficient in vesicles, can be studied.

Continue east across a columnar-jointed, coarse-grained dolerite sheet towards the summit of Pen-y-Gadair. At the distinct step up to the summit, a rhyolitic vitric tuff with lapilli-rich horizons is exposed and the summit itself consists of overlying basic pumice tuff. Stevenson maps all these volcanic rocks and associated sediments as the Lower Division of the Crafnant Volcanic Group, whereas Davies maps them as the Bedded Pyroclastic Series. They are all probably marine.

Descend eastward across exposures of slate and cross the largely unexposed ground to the superbly preserved hill fort on the summit of Pen-y-Gaer. The Roman fort of Canovium can be seen, 3 km away to the northeast, on the western bank of the Afon Conwy. The fort at Pen-y-

Gaer is built on rhyolitic vitric tuffs which are cleaved and columnar-jointed. The tuffs dip 25° towards 095° and have been mapped as members of the Middle Division of the Crafnant Volcanic Group by Stevenson. Davies has mapped them as the Bedded Pyroclastic Series, which is curious, since they are rhyolitic rather than basic.

Take the road from Pen-y-Gaer to Llanbedr-y-cennin. At several points small roadside exposures of grey Ashgillian mudstones can be examined; and small brachiopods and sometimes fragmentary trilobite remains can be obtained.

Return to Tal-y-bont.

Glossary Place Names and Grid References

Aber, 656727
Aberdaron, 173264
Abererch, 397366
Aberglaslyn, 595462
Abersoch, 314281
Afon Aber, 662718–648737
Afon Anafon, 698699–662718
Afon Colwyn, 597522–592481
Afon Conwy, 800542–775795
Afon Crafnant, 754616–782637
Afon Ddu, 712720–678755
Afon Ddwyryd, 382389–361348
Afon Denau, 676628–665605
Afon Dwyfach, 463485–465380
Afon Dwyfor, 540508–479373
Afon Erch, 385445–382353
Afon Glaslyn, 633544–583385
Afon Goch, 545545–552553
Afon Gwyrfai, 568524–453591
Afon Lloer, 663620–665607
Afon Llugwy, 692625–798573
Afon Nant Peris, 647557–599588
Afon Porth Llwyd, 723649–775680
Afon Seiont, 560623–476627
Afon Soch, 248355–314283
Allt Fawr, 385357
Arddu, Yr, 628463
Arenig, 826369

Bangor, 580720
Bangor Mountain, 583720
Bardsey Island, 120220
Beaumaris (Anglesey), 600760
Beddgelert, 590482
Benarth, 788767
Bethesda, 625655
Betws Garmon, 536575
Blackrock Halt, 519378
Blackrock Sands, 525375–540365
Blaenau Dolwyddelan, 700517
Blaenau Ffestiniog, 705455
Blaen-y-nant, 642608
Bodidda, 766757
Bodlondeb, 780780
Bodwrog, 319314
Bont Newydd, 662720
Botwnnog, 262314
Braich y Foel, 563585
Bryn Banog, 576457
Bryn-engan, 718577
Bwlch Ciliau, 620536
Bwlch Coch, 622552
Bwlch Cwm Brwynog, 591557
Bwlch Cwm-cesig, 558572
Bwlch Ddwy Glyder, 652582
Bwlch Goleuni, 696583
Bwlch Gwyn (Anglesey), 483729
Bwlch Gwyn, 557574
Bwlch Mawr, 426478
Bwlch Meillionen, 560476
Bwlch Tryfan, 662589
Bwlch y Brecan, 625616
Bwlch y Cywion, 628608
Bwlch y Ddeufaen, 715717
Bwlch-y-groes, 557597
Bwlch-y-saethau, 615542
Bwlch Yr Eifl, 362454
Bytilith, 224264
Cader Cawrdaf, 405365
Cadnant, 774776
Cae Coch, 565722
Caernarfon, 480625
Canovium, 777703
Capel Curig, 721581
Carmel, 495552
Carn Bach, 285347
Carnedd Dafydd, 663631
Carnedd Llywelyn, 684644
Carnedd y Filiast, 621628
Carneddol, 302332
Carreg Ddu, 149239
Carreg Dinas, 287353
Carreg y Defaid, 342324
Castell, 563495
Castell Cidwm, 550558
Castle Hill, 588725
Cefn Coch, 724746
Cefn-du, 550610
Cefn y Capel, 703583
Cerrig Cochion (Dolgarrog), 725762
Cerrig Cochion (Nant Gwynant), 662510
Cerris, 556712
Ceunant Ciprwth, 515487–531467
Ceunant Mawr, 665536
Ceunant Ty'n-y-Ddol, 672522–703514
Chwilog, 435384
Cilcoed, 425502
Clip-y-Gylfinir, 224285
Clipiau, 417464
Clogwyn Bach, 377360
Clogwyn Du, 647581
Clogwyn du'r Arddu, 600555

Llwytmor, 686695
Llyn Anafon, 697698
Llyn Bochlwyd, 654593
Llyn Bychan, 752593
Llyn Caseg-fraith, 670583
Llyn Cowlyd, 730625
Llyn Cwellyn, 560550
Llyn Cwm-bach, 563409
Llyn Cwmdulyn, 493495
Llyn Cwm Silin, 514507
Llyn Cwm Ystradllyn, 562444
Llyn Dinas, 616495
Llyn du'r Arddu, 601557
Llyn Dwythwch, 570580
Llyn Eigiau, 720650
Llyn Geirionydd, 764610
Llyn Gwynant, 645520
Llyn Idwal, 645595
Llyn Ogwen, 660665
Llyn Padarn, 570615
Llyn Peris, 592592
Llyn-y-Cŵn, 637584
Llynnau Mymbyr, 708575

Madryn, 286362
Maes, Y, 578714
Maes y Gaer, 663726
Maescaradog, 635627
Marchlyn Bach, 607625
Marchlyn Mawr, 616620
Meillionydd, 217293
Minffordd, 600385
Moel Bronmiod, 413456
Moel Caerau, 292355
Moel Cynghorion, 586564
Moel Ddu, 579442
Moel Eilio, 556577
Moel Hebog, 565469
Moel Lefn, 553486
Moel Meirch, 661503
Moel Siabod, 705547
Moel Tryfan, 515562
Moel y Dyniewyd, 612478
Moel y Penmaen, 338386
Moel yr Ogof, 556478
Moelfre, 395447

Moelwyn, 658449
Morfa Nefyn, 287403
Mountain Cottage Quarry, 230347
Mynydd Anelog, 152273
Mynydd Carnguwch, 374429
Mynydd Cilgwyn, 497544
Mynydd Drws-y-coed, 544517
Mynydd Gorllwyn, 575425
Mynydd Gwyddel, 142252
Mynydd Llyndy, 621485
Mynydd Mawr, 539547
Mynydd Penarfynydd, 220265
Mynydd Perfedd, 624619
Mynydd Rhiw, 228294
Mynydd Tal-y-mignedd, 535514
Mynydd Tir-y-cwmwd, 329309
Mynydd-y-Dref (Conwy Mountain), 760778
Mynydd-y-Graig, 228275
Mynytho, 308312

Nanmor, 601460
Nant Braich-y-ddinas, 524497–534483
Nant Ffrancon, 647605–630650
Nant Geuallt, 736587–727577
Nant Peris, 605584
Nant-y-gwryd, 662562–720578
Nant-yr-ogof, 673583–693601
Nantlle, 510535
Nefyn, 307407

Ogof ddu, 513379
Ogwen Bank, 626655
Ogwen Cottage, 651603

Parc, 527586
Parwyd, 155243
Pen Benar, 318282
Pen Llithrig-y-wrâch, 716623
Pen Morfa, 548406
Pen-y-Castell, 723689
Penychain, 436353
Pen-y-cil, 158239
Pen-y-Drum, 771609
Pen-y-Gadair, 739694
Pen-y-Gaer (Clynnog), 429457
Pen-y-Gaer (Llanbedr), 750693

Pen-y-garreg, 426497
Pen-y-groes, 471532
Pen-y-pass, 647556
Pen-yr-allt (Pwllheli), 371354
Pen-yr-allt (Trefriw), 769622
Pen-yr-Helgi-ddu, 698631
Pen-yr-Ole Wen, 656620
Penmaen-bach, 747781
Penmaen-mawr, 715765
Penrhyn Du, 324267
Penrhyn Nefyn, 296410
Penrhyndeudraeth, 612389
Pentre, 639615
Pentre'r-felin, 527397
Pitt's Head, 576515
Plas Dinorwic, 526679
Plas-y-nant, 551562
Pont Cae'r-gors, 576508
Pont Pen-y-benglog, 649605
Pont Tal-y-llyn, 668606
Pont-y-bettws, 534577
Pont y Ceunant, 633643
Pont y Ddwyryd, 370358
Pont-y-gromlech, 629566
Porth Ceiriad, 305247–317248
Porth Dinllaen, 277410–295410
Porth Dinorwic, 525675
Porth Felen, 144249
Porth Llwyd, 768677
Porth Mawr, 322265–317280
Porth Nefyn, 295408–318420
Porth Oer, 165298
Porthmadog, 568388
Pwllheli, 375350

Ravenscrag, 760766
Rhaeadr–fawr, 668700
Rhiw-for-fawr, 514381
Rhiw, Y, 228278
Rhos-las, 495503
Rhyd-ddu, 569528
Rhyd-y-clafdy, 328349
Roman Bridge, 713514
Rowlyn-isaf, 754679

Sarn, 239325
St Mary's Well, 139252

References

Bassett D A 1969 Some of the major structures of early Palaeozoic age in Wales and the Welsh Borderland: an historical essay *The Pre-Cambrian and Lower Palaeozoic Rocks of Wales* ed A Wood (Cardiff: University of Wales Press) pp 67–116

Bassett D A and Walton E K 1959 The Hell's Mouth Grits: Cambrian Greywackes in St Tudwal's Peninsula, North Wales *Q. J. Geol. Soc. Lond.* **116** 95–110

Bassett M C, Owens R M and Rushton A W A 1976 Lower Cambrian fossils from the Hell's Mouth Grits, St Tudwal's Peninsula, North Wales *J. Geol. Soc. Lond.* **132** 623–44

Beavon R V 1963 The succession and structure east of the Glaslyn River, North Wales *Q. J. Geol. Soc. Lond.* **119** 479–512

Bonney T G 1879 On the quartz–felsite and associated rocks at the base of the Cambrian Series in northwestern Caernarvonshire *Q. J. Geol. Soc. Lond.* **35** 309–20

Bromley A V 1969 Acid plutonic igneous activity in the Ordovician of North Wales *The Pre-Cambrian and Lower Palaeozoic Rocks of Wales* ed A Wood (Cardiff: University of Wales Press) pp 387–408

Cattermole P 1969 A preliminary geochemical study of the Mynydd Penarfynydd intrusion, Rhiw igneous complex, south-west Lleyn *The Pre-Cambrian and Lower Palaeozoic Rocks of Wales* ed A Wood (Cardiff: University of Wales Press) pp 435–46

Cattermole P and Jones A 1970 The geology of the area around Mynydd Mawr, Nantlle, Caernarvonshire *Geol. J.* **7** 111–28

Cowie J W, Rushton A W A and Stubblefield C J 1972 A correlation of the Cambrian rocks in the British Isles *Geol. Soc. Lond. Spec. Rep. No.* 2

Crimes T P 1969a The stratigraphy, structure and sedimentology of some of the Pre-Cambrian and Cambro-Ordovician rocks bordering the southern Irish Sea *PhD Thesis* University of Liverpool

—— 1969b Trace fossils from the Cambro-Ordovician rocks of North Wales, UK, and their stratigraphical significance *Geol. J.* **6** 333–8

—— 1970a A facies analysis of the Cambrian of Wales *Palaeogeogr., Palaeoclimatol., Palaeoecol.* **7** 113–70

—— 1970b A facies analysis of the Arenig of Western Lleyn, North Wales *Proc. Geol. Assoc.* **81** 221–40

Dakyns J R and Greenly E 1905 On the probable Péléean origin of the Felsitic Slates of Snowdon and their metamorphism *Geol. Mag.* **42** 541–9

Davies D A B 1936 The Ordovician rocks of the Trefriw district (North Wales) *Q. J. Geol. Soc. Lond.* **92** 62–90

Davies E (ed) 1958 *A Gazetteer of Welsh Place-names* (Cardiff: University of Wales Press)

Davies R A 1969 Geological succession and structure of Cambrian and Ordovician rocks in the northeastern Carneddau *PhD Thesis* University of Wales, Aberystwyth

Dewey J F 1969a Evolution of the Appalachian/Caledonian orogen *Nature* **222** 124–9

—— 1969b Structure and sequence in paratectonic British Caledonides, in *North Atlantic—Geology and Continental Drift* ed M Kay *Mem. Am. Assoc. Petrol. Geol.* **12** 309–35

Diggens J M and Romano M 1968 The Caradoc rocks around Llyn Cowlyd, North Wales *Geol. J.* **6** 31–48

Elles G L 1904 Some graptolite zones in the Arenig rocks of Wales *Geol. Mag.* **41** 199–211

—— 1909 The relation of the Ordovician and Silurian rocks of Conway (North Wales) *Q. J. Geol. Soc. Lond.* **65** 169–94

Evans C D R 1968 Geological succession and structure of the area east of Bethesda *PhD Thesis* University of Wales, Aberystwyth

Fearnsides W G 1910 The Tremadoc Slates and associated rocks of south-east Caernarvonshire *Q. J. Geol. Soc. Lond.* **66** 142–88

Fearnsides W G and Davies W 1944 The geology of Deudraeth: the country between Traeth Mawr and Traeth Bach, Merioneth *Q. J. Geol. Soc. Lond.* **99** 247–76

Fitch F J 1967 Ignimbrite volcanism in North Wales *Bull. Volcanol.* **30** 199–219

Fitch F J, Miller J A, Evans A L, Grasty R L and Meneisy M Y 1969 Isotopic age determinations on rocks from Wales and the Welsh Borders *The Pre-Cambrian and Lower Palaeozoic Rocks of Wales* ed A Wood (Cardiff: University of Wales Press) pp 23–45

Fitton J G and Hughes D J 1970 Volcanism and plate tectonics in the British Ordovician *Earth Planet. Sci. Lett.* **8** 223–8

Francis E H and Howells M F 1973 Transgressive welded ash-flow tuffs among the Ordovician sediments of NE Snowdonia, N Wales *J. Geol. Soc. Lond.* **129** 621–41

George T N 1961 *British Regional Geology: North Wales* 3rd edn (London: HMSO)

Greenly E 1919 Geology of Anglesey *Mem. Geol. Surv. GB* (2 vols)

—— 1928 The Lower Carboniferous rocks of the Menaian region of Caernarvonshire *Q. J. Geol. Soc. Lond.* **84** 382–439

—— 1930 A question of nomenclature *Geol. Mag.* **67** 287

—— 1944 The Cambrian rocks of Arvon *Geol. Mag.* **81** 170–5

—— 1945 The Arvonian rocks of Arvon *Q. J. Geol. Soc. Lond.* **100** 269–87

—— 1946 The geology of the City of Bangor *Proc. Liverpool Geol. Soc.* **19** 105–12
Griffiths D H and Gibb R A 1965 Bouguer gravity anomalies in Wales *Geol. J.* **4** 335–42
Harper J C 1956 The Ordovician succession near Llanystumdwy, Caernarvonshire *Liverpool Manchester Geol. J.* **1** 385–93
Hawkins T R W 1970 Hornblende gabbros and picrites at Rhiw, Caernarvonshire *Geol. J.* **7** 1–24
Helm D G, Roberts B and Simpson A 1963 Polyphase folding in the Caledonides south of the Scottish Highlands *Nature* **200** 1060–2
Hofmann H J 1975 *Bolopora* not a bryozoan, but an Ordovician phosphatic, oncolitic accretion *Geol. Mag.* **112** 523–6
Howell B F and Stubblefield C J 1950 A revision of the fauna of the North Welsh *Conocophyre viola* Beds implying a Lower Cambrian age *Geol. Mag.* **87** 1–16
Howells M F, Leveridge B E and Evans C D R 1973 Ordovician ash-flow tuffs in eastern Snowdonia *Inst. Geol. Sci. Rep. No.* 73/3
Hughes E W 1917 On the geology of the district from Cil-y-Coed to the St Anne's–Llanllyfni Ridge (Caernarvonshire) *Geol. Mag.* **54** 12–25
Jenkins D A and Ball D F 1964 Pumpellyite in Snowdonian soils and rocks *Mineral. Mag.* **33** 1093–6
Lynas B D T 1973 The Cambrian and Ordovician rocks of the Migneint area, North Wales *J. Geol. Soc. Lond.* **129** 481–504
Matley C A 1913 The geology of Bardsey Island, with an appendix on the petrography of Bardsey Island by J S Flett *Q. J. Geol. Soc. Lond.* **69** 514–33
—— 1928 The Pre-Cambrian complex and associated rocks of southwestern Lleyn (Caernarvonshire) *Q. J. Geol. Soc. Lond.* **84** 440–504
—— 1932 The geology of the country around Mynydd Rhiw and Sarn, southwestern Lleyn *Q. J. Geol. Soc. Lond.* **88** 238–73
—— 1938 The geology of the country around Pwllheli, Llanebedrog and Madryn (southwestern Caernarvonshire) *Q. J. Geol. Soc. Lond.* **94** 555–606
Matley C A and Heard A 1930 The geology of the country around Bodfean *Q. J. Geol. Soc. Lond.* **86** 130–68
Matley C A and Smith B 1936 The age of the Sarn granite *Q. J. Geol. Soc. Lond.* **92** 188–200
Miyashiro A 1961 Evolution of metamorphic belts *J. Petrol.* **2** 277–311
Moorbath S and Shackleton R M 1966 Isotopic ages for the Precambrian Mona Complex of Anglesey, North Wales, Great Britain *Earth Planet. Sci. Lett.* **1** 113–7
Morris T O and Fearnsides W G 1926 The stratigraphy and structure of the Cambrian slate belt of Nantlle (Caernarvonshire) *Q. J. Geol. Soc. Lond.* **82** 250–303
Nicholas T C 1915 The geology of the St Tudwal's Peninsula (Caernarvonshire) *Q. J. Geol. Soc. Lond.* **71** 83–143
—— 1916 Notes on the trilobite fauna of the Middle Cambrian of the St Tudwal's Peninsula (Caernarvonshire) *Q. J. Geol. Soc. Lond.* **71** 451–72
Oxburgh E R and Turcotte D L 1971 Origin of paired metamorphic belts and crustal dilation in island arc regions *J. Geophys. Res.* **76** 1315–27
Ramberg H 1970 Model studies in relation to intrusions of plutonic bodies *Mechanism of Igneous Intrusions* (*Geol. J. Sp. Iss.*) ed G Newall and N Rast (Liverpool: Gallery Press) pp 261–86
Ramsay A C 1866 The geology of North Wales *Mem. Geol. Surv. GB* vol 3
Rast N 1961 Mid-Ordovician structures in south-western Snowdonia *Liverpool Manchester Geol. J.* **2** 645–52
—— 1969 The relationship between Ordovician structure and volcanicity in Wales *The Pre-Cambrian and Lower Palaeozoic Rocks of Wales* ed A Wood (Cardiff: University of Wales Press) pp 305–36
Roberts B 1967 Succession and structure in the Llwyd Mawr syncline, Caernarvonshire, North Wales *Geol. J.* **5** 369–90
—— 1969 The Llwyd Mawr ignimbrite and its associated volcanic rocks *The Pre-Cambrian and Lower Palaeozoic Rocks of Wales* ed A Wood (Cardiff: University of Wales Press) pp 337–56
—— 1975 The Cwm Dulyn rhyolite, Snowdonia *Geol. Mag.* **112** 416–8
Roberts B and Siddans A W B 1971 Fabric studies in the Llwyd Mawr ignimbrite, Caernarvonshire, North Wales *Tectonophysics* **12** 283–306
Shackleton R M 1953 The structural evolution of North Wales *Liverpool Manchester Geol. J.* **1** 261–97
—— 1954 The structure and succession of Anglesey and the Lleyn Peninsula *Adv. Sci. Lond.* **XI** (41) 106–8
—— 1956 Notes on the structure and relations of the Pre-Cambrian and Ordovician rocks of south-western Lleyn (Caernarvonshire) *Liverpool Manchester Geol. J.* **1** 400–9
—— 1959 The stratigraphy of the Moel Hebog district between Snowdon and Tremadoc *Liverpool Manchester Geol. J.* **2** 216–51
—— 1969 The Pre-Cambrian of North Wales *The Pre-Cambrian and Lower Palaeozoic Rocks of Wales* ed A Wood (Cardiff: University of Wales Press) 1–22
Skevington D 1969 The classification of the Ordovician system in Wales *The Pre-Cambrian and Lower Palaeozoic Rocks of Wales* ed A Wood (Cardiff: University of Wales Press) pp161–80
Stevenson I P 1971 The Ordovician rocks of the country between Dwygyfylchi and Dolgarrog, Caernarvonshire *Proc. Yorkshire Geol. Soc.* **38** 517–48
Toghill P 1970 A fauna from the Hendre Shales (Llandeilo) of the Mydrim area, Carmarthenshire *Proc. Geol. Soc. Lond.* **1663** 121–9
Tremlett W E 1962 The geology of the Nefyn–Llanaelhaearn area of North Wales *Liverpool Manchester Geol. J.* **3** 157–76
—— 1964 The geology of the Clynnog-fawr district and Gurn Ddu Hills of northeast Lleyn *Geol. J.* **4** 207–23
—— 1965 The geology of the Chwilog area of southeastern Lleyn (Caernarvonshire) *Geol. J.* **4** 435–48
Williams A, Strachan I, Bassett D A, Dean W T, Ingham J K, Wright A D and Whittington H B 1972 A correlation of Ordovician rocks in the British Isles *Geol. Soc. Lond. Spec. Rep. No.* 3
Williams D 1930 The geology of the country between Nant Peris and Nant Ffrancon (Snowdonia) *Q. J. Geol. Soc. Lond.* **86** 191–233
Williams H 1922 The igneous rocks of the Capel Curig district (North Wales) *Proc. Liverpool Geol. Soc.* **13** 166–206
—— 1927 The geology of Snowdon (North Wales) *Q. J. Geol. Soc. Lond.* **83** 346–431

Williams H and Bulman O M B 1931 The geology of the Dolwyddelan Syncline (North Wales) *Q. J. Geol. Soc. Lond.* **87** 425–58

Winkler H G F 1967 *The Petrogenesis of Metamorphic Rocks* 2nd edn (Berlin: Springer)

Wood D S 1969 The base and correlation of the Cambrian rocks of North Wales *The Pre-Cambrian and Lower Palaeozoic Rocks of Wales* ed A Wood (Cardiff: University of Wales Press) pp 47–66

—— 1974 Ophiolites, mélanges, blueschists and ignimbrites: early Caledonian subduction in Wales? *Modern and Ancient Geosynclinal Sedimentation: Soc. Econ. Palaeontol. Mineral. Spec. Publ. No.* 19 ed R H Dott and R H Shaver pp 334–44

Wood D S and Harper J C 1962 Notes on a temporary section in the Ordovician at Conway, North Wales *Liverpoool Manchester Geol. J.* **3** 177–86

Wright J B 1974 The Cwm Dulyn rhyolite, Snowdonia—an extrusive dome? *Geol. Mag.* **111** 444–6

Wynne-Hyghes E 1917 On the geology of the district from Cil-y-Coed to the St Anne's–Llanllyfni Ridge (Caernarvonshire) *Geol. Mag.* **4** 12–25, 75–80

Index